Lecture Notes in Mathematics

Edited by A. Dold and B. Eckmann

463

Hui-Hsiung Kuo

Gaussian Measures in Banach Spaces

Springer-Verlag
Berlin · Heidelberg · New York 1975

Author
Prof. Hui-Hsiung Kuo
University of Virginia
Department of Mathematics
Charlottesville, VA 22903
USA

Library of Congress Cataloging in Publication Data

Kuo, Hui-Hsiung, 1941-
Gaussian measures in Banach spaces.

(Lecture notes in mathematics ; v. 463)
Bibliography: p.
Includes index.
1. Gaussian measures. 2. Banach spaces. I. Title.
II. Series: Lecture notes in mathematics (Berlin) ;
v. 463.
QA3.L28 no. 463 [QA312] 510'.8s [515'.42] 75-16345

AMS Subject Classifications (1970): 28A40

ISBN 3-540-07173-3 Springer-Verlag Berlin · Heidelberg · New York
ISBN 0-387-07173-3 Springer-Verlag New York · Heidelberg · Berlin

Printed in Germany
Offsetdruck: Julius Beltz, Hemsbach/Bergstr.

PREFACE

This monograph is based on the lecture notes of a course entitled "Applications of Measure Theory" given in the Spring of 1974 at the University of Virginia. As the reader can easily see, the material delivered does not cooperate with the course's title very well. Our primary object in this course was to give an introduction to the notion of abstract Wiener space and study some of the related topics. We covered the first two chapters and the first three sections of Chapter III. The last four sections were added in when I rewrote the lecture notes. I deeply regret that in this course we did not discuss in details the recent works of J. Eells, K.D. Elworthy and R. Ramer, among others on the integration on Banach manifolds. I feel that it would be too ambitious to include their works in these notes.

I would like to express my appreciation to Professor Leonard Gross and Professor Kiyosi Ito for their constant encouragement and mathematical influence. The conversations with them have always been a source of inspiration. I would like to thank Tavan Trent for proof-reading parts of the manuscript. My special thanks go to Barbara Smith and Fukuko Kuo for typing the manuscript. The preparation of these notes is partially supported by the National Science Foundation.

H. -H. Kuo

TABLE OF CONTENTS

Chapter I. Gaussian measures in Banach spaces.

The Lebesgue measure plays a fundamental role in the integration theory in $\mathbb{R}^n$. Recall that it is uniquely determined (up to some constant) by the following conditions: (a) it assigns finite values to bounded Borel sets and positive numbers to non-empty open sets (b) it is translation invariant. Mathematically, one may ask the question : Does the Lebesgue measure make sense in infinite dimensional space ? The answer is negative. To make our assertion precise, consider a separable Hilbert space H. Let μ be a Borel measure in H. We require that μ satisfies the above conditions (a) and (b). We want to get a contradiction. Let $\{e_1, e_2, \ldots\ldots\}$ be an orthonormal basis of H. Let B_n be the ball of radius $\frac{1}{2}$ centered at e_n, and B the ball of radius 2 centered at the origin. Then $0 < \mu(B_1) = \mu(B_2) = \mu(B_3) = \ldots\ldots < \infty$. Note that the B_n's are disjoint and contained in B. Therefore, we must have $\mu(B) \geq \sum_n \mu(B_n) = \infty$. This contradicts (a). Observe that the same argument shows the non-existence of μ even if we replace translation invariance by rotation invariance.

Fortunately, the Gaussian measure makes sense in infinite dimensional space. This will be the center of our investigation in this chapter. The Gaussian measure in $\mathbb{R}^n$ is given as follows:

$$p_t(E) = (2\pi t)^{-n/2} \int_E e^{-|x|^2/2t} \, dx, \quad E \in \mathcal{B}(\mathbb{R}^n).$$

Note that p_t is rotation invariant. In the Hilbert space case, we have just seen that p_t can not be rotation invariant. However, it is rotation invariant with respect to the rotations of another Hilbert space which is embedded in the original one. This will be clear later in this chapter.

We will first discuss Borel measures in Hilbert spaces (due essentially to Prohorov [38] , Sazonov [40] and Gross [17]). Then we will discuss the Gaussian measures in Banach spaces (due essentially to Wiener [48; 49] Gross [18]). In order to study Borel measures in Hilbert spaces, we review Hilbert-Schmidt and trace class operators (see, e.g., [13]).

§1. Hilbert-Schmidt and trace class operators.

Let H be a separable Hilbert space with norm $|\cdot| = \sqrt{<\cdot,\cdot>}$. Let A be a linear operator of H.

Theorem 1.1. Let $\{e_n\}$ and $\{d_n\}$ be any two orthonormal bases of H, then

$$\sum_{n=1}^{\infty} |Ae_n|^2 = \sum_{n=1}^{\infty} |Ad_n|^2.$$

Remark. The above theorem says implicity that if $\sum_{n=1}^{\infty} |Ae_n|^2$ is convergent for some $\{e_n\}$, so is for any other $\{d_n\}$. And if $\sum_{n=1}^{\infty} |Ae_n|^2$ is divergent, so is for any other $\{d_n\}$.

Proof. Note that $|Ae_n|^2 = \sum_{m=1}^{\infty} |< Ae_n, d_m>|^2$. Hence we have

$$\sum_n |Ae_n|^2 = \sum_n \sum_m |<Ae_n, d_m>|^2 = \sum_n \sum_m |<e_n, A^*d_m>|^2$$

$$= \sum_m \sum_n |<e_n, A^*d_m>|^2 = \sum_m |A^* d_m|^2.$$

The above identity is true for any $\{e_n\}$ and $\{d_m\}$. Thus if we put $d_m = e_m$; we have for any orthonormal basis $\{d_n\}$,

$$\sum_m |Ad_m|^2 = \sum_m |A^*d_m|^2.$$

Putting this to the above identity, we have $\sum_n |Ae_n|^2 =$

$$\sum_m |A^*d_m|^2 = \sum_m |Ad_m|^2. \qquad \#$$

Definition 1.1. A linear operator A of H is called a Hilbert-Schmidt operator if, for some orthonormal basis $\{e_n\}$ of H, $\sum_{n=1}^{\infty} |Ae_n|^2 < \infty$. The Hilbert-Schmidt norm of A is defined as follows:

$$||A||_2 = \left(\sum_{n=1}^{\infty} |Ae_n|^2 \right)^{\frac{1}{2}}.$$

Remark. Note that $||A||_2$ does not depend on the choice of $\{e_n\}$ by Theorem 1.1.

Theorem 1.2. (a) $||A^*||_2 = ||A||_2$,

(b) $||\alpha A||_2 = |\alpha| \, ||A||_2$, α: scalar,

(c) $||A + B||_2 \leq ||A||_2 + ||B||_2$,

(d) $||A|| \leq ||A||_2$, where $||A|| = \sup_{x \neq 0} \frac{|Ax|}{|x|}$,

(e) $||AB||_2 \leq ||A|| \, ||B||_2$, $||AB||_2 \leq ||A||_2 \, ||B||$,

Remark. (a) says implicitly that if A is a Hilbert-Schmidt operator then its adjoint operator A* is also Hilbert-Schmidt. Similar explanation should be applied to the other statements.

Proof. (a) follows from the proof of Theorem 1.1.

(b) is obvious.

(c) follows from $|(A + B)\,x| \le |A\,x| + |B\,x|$ and the Minkowski inequality:

$$\left(\sum_{k=1}^{\infty} |\alpha_k + \beta_k|^2\right)^{\frac{1}{2}} \le \left(\sum_{k=1}^{\infty} |\alpha_k|^2\right)^{\frac{1}{2}} + \left(\sum_{k=1}^{\infty} |\beta_k|^2\right)^{\frac{1}{2}}.$$

Now, $|A\,x|^2 = \sum_{n=1}^{\infty} |\langle Ax, e_n\rangle|^2 = \sum_{n=1}^{\infty} |\langle x, A^*e_n\rangle|^2$

$$\le \sum_{n=1}^{\infty} |x|^2 |A^*e_n|^2$$

$$= |x|^2 \sum_{n=1}^{\infty} |A^*e_n|^2 = |x|^2 ||A^*||_2^2 = |x|^2 ||A||_2^2.$$

Therefore, $|A\,x| \le |x|\ ||A||_2$ and this gives (d).

Finally, (e) can be shown as follows:

$$||AB||_2^2 = \sum_{n=1}^{\infty} |ABe_n|^2 \le \sum_{n=1}^{\infty} ||A||^2 |Be_n|^2 = ||A||^2 \sum_{n=1}^{\infty} |Be_n|^2$$

$$= ||A||^2 ||B||_2^2, \text{ so } ||AB||_2 \le ||A||\ ||B||_2.$$

Moreover, $||AB||_2 = ||(AB)^*||_2 = ||B^*A^*||_2 \le ||B^*||\ ||A^*||_2$

$$= ||B||\ ||A||_2.$$

<u>Notation</u>. $L_{(2)}(H)$ denotes the collection of Hilbert-Schmidt operators of H. $L(H)$ denotes the collection of bounded linear operators of H. By Theorem 1.2(d), $L_{(2)}(H) \subset L(H)$. If H is finite dimensional, then $L_{(2)}(H) = L(H)$. But if H is ∞-dimensional, then $L_{(2)}(H) \neq L(H)$, e.g. the identity operator I of H is in $L(H)$, but not in $L_{(2)}(H)$.

<u>Definition 1.2</u>. Let A and B be in $L_{(2)}(H)$. Define the <u>Hilbert-Schmidt inner product</u> $\langle\langle A,B\rangle\rangle$ of A and B as follows:

$$\langle\langle A,B\rangle\rangle = \sum_{n=1}^{\infty} \langle Ae_n, Be_n\rangle,$$

where $\{e_n\}$ is an orthonormal basis of H.

Remark. The above series converges absolutely, because $2| \langle Ae_n,Be_n\rangle | \leq |Ae_n|^2 + |Be_n|^2$. Moreover, using the same arguement in the proof of Theorem 1.1. we can easily see that $\langle\langle A,B\rangle\rangle$ is well-defined.

Theorem 1.3. $L_{(2)}(H)$ with the inner product $\langle\langle\cdot,\cdot\rangle\rangle$ is a Hilbert space.

Proof. Theorem 1.2 (b) and (c) show that $L_{(2)}(H)$ is a vector space. Clearly, $\langle\langle A,A\rangle\rangle = ||A||_2^2$. We show the completeness of $L_{(2)}(H)$. Let $\{A_n\}$ be a Cauchy sequence in $L_{(2)}(H)$. Because of Theorem 1.2 (d) $\{A_n\}$ is also a Cauchy sequence in $L(H)$. Recall that $L(H)$ is a Banach space with the operator norm. Therefore, there exists $A \in L(H)$ such that $\lim_{n\to\infty} ||A_n - A|| = 0$. We have to prove that $A \in L_{(2)}(H)$ and $\lim_{n\to\infty} ||A_n-A||_2 = 0$. Let $\varepsilon > 0$, then $||A_n - A_m||_2 < \varepsilon$ for sufficiently large n and m. Now,

$$\sum_{k=1}^{s} |(A_n - A_m)e_k|^2 \leq ||A_n - A_m||_2^2 < \varepsilon^2$$

for sufficiently large m and n, and any s. Letting $m \to \infty$ and noting that $\lim_{n\to\infty} ||A_n - A|| = 0$, we have

$$\sum_{k=1}^{s} |(A_n - A)e_k|^2 \leq \varepsilon^2$$

for sufficiently large n and any s. Letting $s \to \infty$, we have

$$\sum_{k=1}^{\infty} |(A_n - A)e_k|^2 \leq \varepsilon^2 < \infty, \text{ n sufficiently large.}$$

Therefore $A_n - A \in L_{(2)}(H)$ and hence $A = A_n + (A - A_n) \in L_{(2)}(H)$. Moreover, $||A_n - A||_2 \leq \varepsilon$ for sufficiently large n. Hence $\lim_{n\to\infty} ||A_n - A||_2 = 0$. #

Example 1. Let A be the operator of ℓ_2 given by

$$A(a_1,a_2,\ldots,a_n,\ldots) = (a_1, \frac{a_2}{2},\ldots,\frac{a_n}{n},\ldots).$$

Then A is a Hilbert-Schmidt operator of ℓ_2, and $||A||_2 = \sqrt{\sum_{n=1}^{\infty} \frac{1}{n^2}}$.

Example 2. Let $A : \ell_2 \to \ell_2$ be given by $A(a_1,a_2,\ldots,a_n,\ldots) = (\alpha_1 a_1, \alpha_2 a_n, \ldots, \alpha_n a_n \ldots)$. Then A is a Hilbert-Schmidt operator if and only if $\sum_{n=1}^{\infty} |\alpha_n|^2 < \infty$.

Example 3. (Hilbert-Schmidt integral operator). Let $k \in L^2(\mathbb{R}^2)$. Define an operator $K : L^2(\mathbb{R}) \to L^2(\mathbb{R})$ by

$$Kf(s) = \int_{-\infty}^{\infty} k(s,t)f(t)dt.$$

Then K is Hilbert-Schmidt and $||K||_2 = (\int_{\mathbb{R}^2} \int |k(s,t)|^2 \, dsdt)^{1/2}$.

Of course, we have to make sure that $Kf \in L^2(\mathbb{R})$ whenever $f \in L^2(\mathbb{R})$.

$$\begin{aligned}\int_{-\infty}^{\infty} |Kf(s)|^2 \, ds &= \int_{-\infty}^{\infty} |\int_{-\infty}^{\infty} k(s,t)f(t)dt|^2 \, ds \\ &\leq \int_{-\infty}^{\infty} (\int_{-\infty}^{\infty} |k(s,t)|^2 \, dt)(\int_{-\infty}^{\infty} |f(t)|^2 \, dt) \, ds \\ &= |f|^2 \int_{-\infty}^{\infty} \int_{-\infty}^{\infty} |k(s,t)|^2 \, dt \, ds = |f|^2 ||k||_2^2.\end{aligned}$$

Hence, we have shown not only that K maps $L^2(\mathbb{R})$ into itself, but also that K is a bounded operator and $||K|| \leq ||k||_2$.

To show that K is actually a Hilbert-Schmidt operator, we have to use Fubini's theorem. Note that $\int_{-\infty}^{\infty} \int_{-\infty}^{\infty} |\overline{k(s,t)}|^2 \, ds \, dt = \int_{-\infty}^{\infty} \int_{-\infty}^{\infty} |k(s,t)|^2 \, ds \, dt < \infty$, hence by Fubini's theorem,

$$\int_{-\infty}^{\infty} |\overline{k(s,t)}|^2 \, dt < \infty \quad \text{a.e.s.}$$

i.e. $\overline{k(s,\cdot)} \in L^2(\mathbb{R})$ a.e.s.

Let $\{e_n\}$ be an orthonormal basis of $L^2(\mathbb{R})$. Then for a.e.s,

$$\int_{-\infty}^{\infty} |\overline{k(s,t)}|^2\, dt = |\overline{k(s,\cdot)}|^2 = \sum_{n=1}^{\infty} |\langle \overline{k(s,\cdot)}, e_n(\cdot)\rangle|^2$$

$$= \sum_{n=1}^{\infty} \left|\int_{-\infty}^{\infty} \overline{k(s,t)}\ \overline{e_n(t)}\, dt\right|^2.$$

Hence,

$$\int_{-\infty}^{\infty}\int_{-\infty}^{\infty} |\overline{k(s,t)}|^2 dt\, ds = \int_{-\infty}^{\infty} \sum_{n=1}^{\infty} \left|\int_{-\infty}^{\infty} \overline{k(s,t)}\ \overline{e_n(t)}\, dt\right|^2 ds,$$

$$= \sum_{n=1}^{\infty} \int_{-\infty}^{\infty} \left|\int_{-\infty}^{\infty} \overline{k(s,t)}\ \overline{e_n(t)}\, dt\right|^2 ds$$

(by monotone convergence theorem).

Therefore, $\sum_n |Ke_n|^2 = \sum_n \int_{-\infty}^{\infty} |Ke_n(s)|^2\, ds = \sum_n \int_{-\infty}^{\infty} \left|\int_{-\infty}^{\infty} k(s,t) e_n(t) dt\right|^2 ds$

$$= \sum_n \int_{-\infty}^{\infty} \left|\int_{-\infty}^{\infty} \overline{k(s,t)}\,\overline{e_n(t)} dt\right|^2 ds = \int_{-\infty}^{\infty}\int_{-\infty}^{\infty} |\overline{k(s,t)}|^2\, dt\, ds$$

$$= \int_{-\infty}^{\infty}\int_{-\infty}^{\infty} |k(s,t)|^2\, dt\, ds = ||k||_2^2 < \infty. \qquad \#$$

We now discuss the trace class operators of a separable Hilbert space H. In order to do this we have to study the polar decomposition of operators. Recall that a complex number z can be written as follows: $z = e^{i\theta} a$, where $|e^{i\theta}| = 1$ and $a > 0$. The analogue of $e^{i\theta}$ is an isometry operator and analogue of a is a positive operator.

<u>Definition 1.3</u>. An operator of H is called <u>compact</u> if it takes any bounded subset of H into a set whose closure is compact.

<u>Exercise 1</u>. Show that a compact operator is a bounded operator. Show also that if dim (H) = ∞ then the identity operator is not compact.

Exercise 2. (a) Show that a projection operator is compact if and only if the range is of finite dimensional. (b) If A is a compact operator and B is a bounded operator then both AB and BA are compact operators.

Exercise 3. Suppose $\{e_n\}$ is an orthonormal basis of H and $\{\beta_n\}$ is a sequence of numbers converging to zero. Define $Tx = \sum_{n=1}^{\infty} \beta_n \langle x,e_n\rangle e_n$. Show that T is a compact operator.

Theorem 1.4. (Polar decomposition). Let A be a compact operator of H. Then A can be written in the form A = UT, where T is a positive definite compact operator of H and U is an isometry mapping the range of T into H.

Remarks. (a) Of course U can be extended to the closure of T(H) by continuity. Usually, U is extended to H linearly by defining U to be 0 in the orthogonal complement of $\overline{T(H)}$. With this U, UT is called the polar decomposition of A.

(b) The point in this theorem is that we have spectral representation for T, and we can define various things for A in terms of this spectral representation

(c) T is very often denoted by $(A^*A)^{\frac{1}{2}}$.

(d) The proof of this theorem depends on the following spectral representation which we will take for granted. (See, e.g. [28].)

Theorem 1.5. If A is a self-adjoint compact operator, then there exists an orthonormal basis $\{e_n\}$ of H such that

$$Ax = \sum_{n=1}^{\infty} \lambda_n \langle x,e_n\rangle e_n,$$

where λ_n's are real numbers and $\lambda_n \to 0$ as $n \to \infty$.

Remark: λ_n's are called eigenvalues and e_n's are called eigenvectors. Note that when A is positive definite, then $\lambda_n \geq 0$.

Proof of Theorem 1.4. Consider the operator $B = A^*A$. B is a compact operator by Exercise 2(b). Moreover, B is positive definite because $\langle Bx,x\rangle = \langle A^*Ax,x\rangle = \langle Ax,Ax\rangle = |Ax|^2 \geq 0$. Therefore, by Theorem 1.5.

$$Bx = \sum_{n=1}^{\infty} \lambda_n \langle x,e_n\rangle e_n,$$

where $\{e_n\}$ is an orthonormal basis of H and $\lambda_n \geq 0$ and $\lambda_n \to 0$ as $n \to \infty$. Define

$$Tx = \sum_{n=1}^{\infty} \sqrt{\lambda_n} \langle x,e_n\rangle e_n.$$

By Exercise 3, T is a compact operator. Obviously, T is positive definite.

Now, define an operator U in the range T(H) of T by

$$U(Tx) = Ax, \quad x \in H.$$

Observe that $|Ax|^2 = \langle Bx,x\rangle = |Tx|^2$. Hence $Tx = 0$ implies $Ax = 0$. Thus U is well-defined and we have $A = UT$. Moreover,

$$|U(Tx)|^2 = |Ax|^2 = |Tx|^2.$$

Therefore, we have $|U(Tx)| = |Tx|$ and so U is an isometry in T(H).

#

Definition 1.4. A compact operator A of H is called a trace class operator if $\sum_{n=1}^{\infty} \lambda_n < \infty$, where λ_n's are the eigenvalues of

$(A^*A)^{\frac{1}{2}}$.

<u>Exercise 4</u>. Prove that a Hilbert-Schmidt operator is compact. Furthermore, A is a Hilbert-Schmidt operator iff $\sum_{n=1}^{\infty} \lambda_n^2 < \infty$, where λ_n's are the eigenvalues of $(A^*A)^{\frac{1}{2}}$. Prove also that in this case $||A||_2 = \left(\sum_{n=1}^{\infty} \lambda_n^2 \right)^{\frac{1}{2}}$.

<u>Notation</u>. $L_{(1)}(H)$ denotes the collection of trace class operators of H. If $A \in L_{(1)}(H)$, define the <u>trace class norm</u> of A by

$$||A||_1 = \sum_{n=1}^{\infty} \lambda_n .$$

<u>Definition 1.5</u>. If $A \in L_{(1)}(H)$, the <u>trace</u> of A is defined as follows:

trace $A = \sum_{n=1}^{\infty} \langle A e_n, e_n \rangle$, where $\{e_n\}$ is any orthonormal basis of H.

Remarks. It is easy to see that trace A is independent of the choice of $\{e_n\}$. Furthermore, let $\{e_n\}$ be the eigenvectors of $(A^*A)^{\frac{1}{2}}$ and let α_n be the corresponding eigen values. Then

$$\sum_{n=1}^{\infty} |\langle Ae_n, e_n \rangle| = \sum_{n=1}^{\infty} |\langle U T e_n, e_n \rangle| = \sum_{n=1}^{\infty} \alpha_n |\langle U e_n, e_n \rangle| \leq \sum_{n=1}^{\infty} \alpha_n .$$

Hence $\sum_{n=1}^{\infty} \langle A e_n, e_n \rangle$ is absolutely convergent.

<u>Theorem 1.6</u>. (a) $||\alpha A||_1 \leq |\alpha| \, ||A||_1$, $\alpha \in \mathbb{C}$,

(b) $||A + B||_1 \leq ||A||_1 + ||B||_1$,

(c) $||A|| \leq ||A||_1$,

(d) If $A, B \in L_{(2)}(H)$ then $AB \in L_{(1)}(H)$ and

$$||AB||_1 \leq ||A||_2 ||B||_2 ,$$

(e) $||A||_2 \leq ||A||_1$,

(f) $||AB||_1 \leq ||A||\ ||B||_1$, $||AB||_1 \leq ||A||_1 ||B||$,

(g) $||A^*||_1 = ||A||_1$.

Proof.

(a) follows from $\alpha A = e^{i\theta}\rho\, UT = (e^{i\theta}U)(\rho T)$, where $\alpha = \rho e^{i\theta}$ and $A = UT$. Note that $e^{i\theta}U$ is an isometry and ρT is a positive definite compact operator.

(b) is left as Exercise 5.

(c) follows from this: $|Ax| = |UT\,x| = |Tx| =$

$$\left|\sum_{n=1}^{\infty} \alpha_n < x,e_n > e_n\right| \leq \sum_{n=1}^{\infty} \alpha_n | < x,e_n > ||e_n|$$

$$\leq \sum_{n=1}^{\infty} \alpha_n |x| = |x| \sum_{n=1}^{\infty} \alpha_n = |x|\,||A||_1.$$

(d) can be shown as follows: let A and B be two Hilbert-Schmidt operators, and let $AB = UT$ be the polar decomposition of AB. Let $\{\alpha_n\}$ be the eigenvalues of T, and $\{e_n\}$ the corresponding eigenvectors.

Then $\alpha_n = <Te_n,e_n> = <U^*UTe_n,e_n>$

$= <UTe_n, Ue_n> = <ABe_n,Ue_n>$

$= <Be_n,A^*Ue_n>$.

Note that $|Ue_n| = 1$ and $<Ue_n,Ue_m> = <U^*Ue_n,e_m>$ $= <e_n,e_m> = \delta_{nm}$, hence $\{Ue_n\}$ is orthonormal (may not be a basis). Hence $\sum_n |A^*Ue_n|^2 \leq ||A^*||_2^2 = ||A||_2^2$. Therefore,

$$||AB||_1 = \sum_{n=1}^{\infty} \alpha_n = \sum_{n=1}^{\infty} < Be_n, A^*Ue_n >$$

$$\leq \left(\sum_{n=1}^{\infty} |Be_n|^2\right)^{\frac{1}{2}} \left(\sum_{n=1}^{\infty} |A^*Ue_n|^2\right)^{\frac{1}{2}}$$

$$\leq ||B||_2 \, ||A||_2.$$

(e) follows from Exercise 4 and the fact that $\sum_{n=1}^{\infty} \lambda_n^2 \leq (\sum_{n=1}^{\infty} \lambda_n)^2$ for nonnegative λ_n.

(f) Suppose $B \in L_{(1)}(H)$ and let $B = UT$ be the polar decomposition of B. Suppose

$Tx = \Sigma\alpha_n < x,e_n > e_n, \quad \alpha_n \geq 0.$

Define $\sqrt{T}\, x = \Sigma\sqrt{\alpha_n} < x,e_n > e_n.$

Clearly, $\sqrt{T} \in L_{(2)}(H)$ and $||\sqrt{T}\,||_2^2 = \Sigma\alpha_n = ||B||_1.$

Hence, by (d), $AB = AUT = (AU\sqrt{T})\sqrt{T} \in L_{(1)}(H)$ because $\sqrt{T} \in L_{(2)}(H)$ and $AU\sqrt{T} \in L_{(2)}(H)$. Moreover, $||AB||_1 \leq ||AU\sqrt{T}||_2 ||\sqrt{T}||_2 \leq ||A|| \; ||U|| \; ||\sqrt{T}||_2^2 = ||A|| \; ||B||_1$. Second half of (f) follows from the first half of (f) and (g).

(g) Let $A \in L_{(1)}(H)$ and $A = UT$ be its polar decomposition. Define $\sqrt{T}$ as above. Then $A = (U\sqrt{T})\sqrt{T}$. Hence $A^* = (\sqrt{T})^*(U\sqrt{T})^* = \sqrt{T}(U\sqrt{T})^*$. Therefore,

$$||A^*||_1 \leq ||\sqrt{T}||_2 ||(U\sqrt{T})^*||_2 = ||\sqrt{T}||_2 ||U\sqrt{T}||_2$$

$$\leq ||\sqrt{T}||_2 ||U|| \; ||\sqrt{T}||_2 = ||\sqrt{T}||_2^2 = ||A||_1.$$

We have thus proved that $||A^*||_1 \leq ||A||_1$ for all $A \in L_{(1)}(H)$. Hence $||A||_1 = ||(A^*)^*||_1 \leq ||A^*||_1$. So we must have $||A^*||_1 = ||A||_1$. #

Incidently, it follows from $A = (U\sqrt{T})\sqrt{T}$ that we have the following

Corollary 1.1. Any trace class operator can be written as a product of two Hilbert-Schmidt operators.

The following theorem is left as Exercise 6.

Theorem 1.7. $L_{(1)}(H)$ is a Banach space with the trace class norm.

Remarks. (a) It follows from Theorem 1.2 (d) and Theorem 1.6 (e) that $||A|| \leq ||A||_2 \leq ||A||_1$. Hence we have the relation

$$L_{(1)}(h) \subset L_{(2)}(H) \subset L(H).$$

Note that if dim H = ∞ the operator A in Example 1 is not a trace class operator. Hence the first inclusion is proper. We saw before that the second inclusion is also proper.

(b) Let $K(H)$ be the collection of compact operators of H. It can be shown that $K(H)$ is a closed subspace of $L(H)$. Hence $K(H)$ is a Banach space with the operator norm. We have

$$L_{(1)}(H) \subset L_{(2)}(H) \subset K(H) \subset L(H).$$

Note that if dim H = ∞, Example 2 shows that the second inclusion is proper. Exercise 1 shows that the last inclusion is also proper.

(c) An operator A is called degenerate if A(H) is finite dimensional. Let $\mathcal{D}$ denote the collection of degenerate operators. Obviously, $\mathcal{D}$ is a subspace of $L(H)$. We have the following.

$\mathcal{D} \subset \mathcal{L}_{(1)}(H) \subset \mathcal{L}_{(2)}(H) \subset \mathcal{K}(H) \subset \mathcal{L}(H)$.

In fact, it can be shown that :

$\mathcal{L}_{(1)}(H)$ = the completion of $\mathcal{D}$ with respect to $||\cdot||_1$.

$\mathcal{L}_{(2)}(H)$ = the completion of $\mathcal{D}$ with respect to $||\cdot||_2$.

$\mathcal{K}(H)$ = the completion of $\mathcal{D}$ with respect to $||\cdot||$.

(d) $\mathcal{L}_{(1)}(H), \mathcal{L}_{(2)}(H)$ and $\mathcal{K}(H)$ will play important roles in the integration theory in ∞-dimensional space.

(e) Evidently if dim $H < \infty$ then

$$\mathcal{D} = \mathcal{L}_{(1)}(H) = \mathcal{L}_{(2)}(H) = \mathcal{K}(H) = \mathcal{L}(H).$$

(f) $\mathcal{K}(H)^* = \mathcal{L}_{(1)}(H)$.

§2. Borel measures in a Hilbert space.

Let H be a real separable Hilbert space. $\mathcal{B}$ will denote the Borel field of H, i.e. the σ-field generated by the open subsets of H. A Borel measure in H is by definition a measure defined in $(H, \mathcal{B})$.

Example 1. The Dirac measure δ_{x_0} at a fixed point x_0 in H, i.e.

$$\delta_{x_0}(E) = \begin{cases} 1 & \text{if } x_0 \in E \\ 0 & \text{if } x_0 \notin E, \end{cases} \qquad E \in \mathcal{B}.$$

Example 2. Let μ be a Borel measure in $\mathbb{R}$. Let e be a unit vector of H, the induced measure $\tilde{\mu}_e$ of μ in the direction e is defined to be the Borel measure in H as follows:

$$\tilde{\mu}_e(E) = \mu(\zeta(E \cap [e])),$$

where [e] denotes the span of e, and $\zeta : [e] \longrightarrow \mathbb{R}$ is given by $\zeta(\alpha e) = \alpha$.

Example 3. Let μ be a Borel measure in $\mathbb{R}^n$. Let P be an orthogonal projection of H with dim P(H) = n and let ξ be an identification of $PH \to \mathbb{R}^n$. The induced measure $\tilde{\mu}$ in H (depending on P and on ξ) is defined by

$$\tilde{\mu}\ (E) = \mu(\xi(E \cap PH)).$$

Definition 2.1. Let μ be a Borel measure in H. The covariance operator S_μ of μ is defined by

$$\langle S_\mu x,y\rangle = \int_H \langle x,z\rangle\langle y,z\rangle\ \mu(dz). \quad x,y \in H.$$

Remark . S_μ may not exist. If S_μ exists, it is positive definite and self-adjoint.

Example 4. $S_{\delta_{x_0}} x = \langle x,x_0\rangle x_0$ because $\langle S_{\delta_{x_0}} x,y\rangle =$

$\langle x,x_0\rangle \langle y,x_0\rangle = \langle\langle x,x_0\rangle x_0,y\rangle$.

Example 5. $S_{\tilde\mu_e} x = \left(\int_{\mathbb{R}} t^2 \mu(dt)\right) \langle x,e\rangle e$ because :

$$\langle S_{\tilde\mu_e} x,y\rangle = \int_H \langle x,z\rangle \langle y,z\rangle \tilde\mu_e(dz) = \int_{[e]} \langle x,z\rangle \langle y,z\rangle \tilde\mu_e(dz)$$

$$= \langle x,e\rangle\langle y,e\rangle \int_{[e]} \langle z,e\rangle^2 \tilde\mu_e(dz) = \langle x,e\rangle \langle y,e\rangle \int_{\mathbb{R}} t^2 \mu(dt).$$

Remark. Suppose $\int_{\mathbb{R}} t^2 \mu(dt) = 1$. Then $S_{\delta_e} = S_{\tilde\mu_e}$ even though δ_e and $\tilde\mu_e$ are two different measures. Hence μ is not uniquely determined by its covariance operator. However, we will see later on that a Gaussian measure of mean zero is completely determined by its covariance operator.

For the sake of convenience, we introduce the following definition.

Definition 2.2. An operator is called an S-operator of H if it is in $L_{(1)}(H)$, positive definite and self-adjoint. $\mathcal{S}$ denotes the collection of $\mathcal{S}$-operators of H.

Remark. Obviously $\mathcal{S}$ is not a vector space. $\mathcal{S}$ will be regarded just as a set. Note that if $A \in \mathcal{S}$ then $||A||_1 =$ trace A.

Theorem 2.1. $\int_H |x|^2 \mu(dx) < \infty$ if and only if $S_\mu \in \mathcal{S}$.

In fact, trace $S_\mu = \int_H |x|^2 \mu(dx)$

Proof.

Sufficiency : Let $\{e_n\}$ be an orthonormal basis of H.

By monotone convergence theorem,

$$\int_H |x|^2 \mu(dx) = \lim_{n\to\infty} \int_H [\langle x,e_1\rangle^2 + \cdots + \langle x,e_n\rangle^2] \mu(dx).$$

But $\int_H \langle x,e_j\rangle^2 \mu(dx) = \langle S_\mu e_j, e_j\rangle$.

$$\text{Hence } \int_H |x|^2 \mu(dx) = \lim_{n\to\infty} \sum_{j=1}^{n} \langle S_\mu e_j, e_j\rangle$$

$$= \sum_{j=1}^{\infty} \langle S_\mu e_j, e_j\rangle$$

$$= \text{trace } S_\mu < \infty.$$

Necessity : First of all, we have to show the existence of the covariance operator. Clearly, $|\langle x,z\rangle \langle y,z\rangle| \leq |x|\ |y| \cdot |z|^2$. Hence

$$\left| \int_H \langle x,z\rangle \langle y,z\rangle \mu(dz)\right| \leq |x||y| \int_H |z|^2 \mu(dz).$$

Therefore, the bilinear form $\int_H \langle x,z\rangle \langle y,z\rangle \mu(dz)$ is continuous. Hence there exists $S_\mu \in L(H)$ such that

$$\langle S_\mu x,y\rangle = \int_H \langle x,z\rangle \langle y,z\rangle \mu(dz).$$

Obviously, S_μ is self-adjoint and positive definite. To show that $S_\mu \in S$, it is sufficient to show that if $\{e_n\}$ is an orthonormal basis of H then the series

$$\sum_{n=1}^{\infty} \langle S_\mu e_n, e_n\rangle \text{ is convergent.}$$

$$\text{But, } \sum_{n=1}^{\infty} \langle S_\mu e_n, e_n\rangle = \sum_{n=1}^{\infty} \int_H \langle x,e_n\rangle^2 \mu(dx)$$

$$= \int_H \sum_{n=1}^{\infty} \langle x, e_n \rangle^2 \mu(dx) \quad \text{(monotone convergence theorem)}$$

$$= \int_H |x|^2 \mu(dx).$$

Hence we have not only shown that $S_\mu \in S$, but also proved that

$$\text{trace } S_\mu = \int_H |x|^2 \mu(dx). \qquad \#$$

Definition 2.3. Let μ be a Borel measure in H, the mean of μ is an element $m_\mu \in H$ such that

$$\langle m_\mu, x \rangle = \int_H \langle z, x \rangle \mu(dz), \quad x \in H.$$

Remark. In general m_μ does not exist. However, if $\int_H |x| \mu(dx) < \infty$ then m_μ exists and $|m_\mu| \leq \int_H |x| \mu(dx)$.

Example 6. Let μ be a Borel measure in $\mathbb{R}$ such that $\int_{\mathbb{R}} t^2 \mu(dt) < \infty$, but $\int_{\mathbb{R}} t\mu(dt) = \infty$. Then $S_{\tilde{\mu}_e}$ exists, but $m_{\tilde{\mu}_e}$ does not exist.

Exercise 7. Construct μ satisfying Example 6. Construct a Borel measure ν in $\mathbb{R}$ such that $S_{\tilde{\nu}_e}$ does not exist, but $m_{\tilde{\nu}_e}$ exists.

Example 7. Let $\{e_n\}$ be an orthonormal basis of H. Let $f_n = \beta_n e_n, \beta_n > 0$, $n=1,2\cdots$. Let $\alpha_n > 0$ and $\Sigma \alpha_n < \infty$. Define a Borel measure μ in H to be the point mass by putting $\mu(\{\beta_n e_n\}) = \alpha_n$, $n = 1,2,\cdots$. Suppose that $\Sigma \alpha_n \beta_n^2 = \infty$ and $\alpha_n \beta_n^2 \to 0$ as $n \to \infty$ (e.g. $\alpha_n = \frac{1}{n^2}$, $\beta_n = \sqrt{n}$).

Then S_μ exists and is given by

$$S_\mu x = \sum_n \alpha_n \beta_n^2 \langle x, e_n \rangle e_n.$$

Obviously, $S_\mu \in K(H)$ but $S_\mu \notin S$.

Exercise 8. Construct μ such that $S_\mu \notin K(H)$, but of course, $S_\mu \in L(H)$. Construct μ such that $S_\mu = I$, the identity operator of H.

Definition 2.4. A function ϕ from H into $\mathbb{C}$ is called a positive definite functional if for any $x_1, x_2, \cdots, x_n \in H$, $n = 1, 2, \cdots$ and any numbers $c_1, c_2, \cdots, c_n$ in $\mathbb{C}$ then

$$\sum_{j,k=1}^{n} c_j \, \phi(x_j - x_k) \bar{c}_k \geq 0.$$

Definition 2.5. The characteristic functional $\hat{\mu}$ of a Borel measure μ in H is defined by

$$\hat{\mu}(x) = \int_H e^{i\langle x, y \rangle} \mu(dy), \quad x \in H.$$

Remark. In general $\hat{\mu}$ may not exist. But if $\mu(H) < \infty$ then $\hat{\mu}$ exists and $|\hat{\mu}(x)| \leq \mu(H)$ for all $x \in H$.

Exercise 9. If μ is a probability measure (i.e. $\mu(H) = 1$) then $\hat{\mu}$ is a positive definite functional and $\hat{\mu}(0) = 1$.

Let μ be a probability measure in H and let ϕ be the characteristic functional of μ. Let us see how smooth ϕ can be. We show below that ϕ is uniformly continuous. Let $\varepsilon > 0$ be given. Then there exists $r > 0$ such that $\mu(S_r) > 1 - \frac{\varepsilon}{4}$, where $S_r = \{x; |x| \leq r\}$. Then

$$|\phi(x) - \phi(y)| \leq \int_H |e^{i\langle x,z\rangle} - e^{i\langle y,z\rangle}|\,\mu(dz)$$

$$= \int_{S_r} + \int_{S_r^c} \quad (S_r^c = \text{complement of } S_r)$$

$$\leq \int_{S_r} |\langle x,z\rangle - \langle y,z\rangle|\,\mu(dz) + 2\cdot\frac{\varepsilon}{4}$$

$$\leq r\,|x - y| + \frac{\varepsilon}{2}.$$

Choose $\delta > 0$ such that $\delta < \frac{\varepsilon}{2r}$. Then

$|x - y| < \delta$ implies $|\phi(x) - \phi(y)| < \varepsilon$.

Hence ϕ is uniformly continuous.

But, if dim $H = \infty$ we can show something more for ϕ. This is the essential conclusion in the following theorem.

__Theorem 2.2.__ (Prohorov) A functional ϕ in H is the characteristic functional of a probability measure in H if and only if (a) $\phi(0) = 1$ and ϕ is positive definite, and (b) for every $\varepsilon > 0$, $\exists\ S_\varepsilon \in \mathcal{S}$ such that

$$1 - \operatorname{Re}\phi(x) \leq \langle S_\varepsilon x,x\rangle + \varepsilon, \text{ for all } x \in H.$$

Remark. (b) implies, in particular, that when $\langle S_\varepsilon x,x\rangle$ is small, then $1 - \operatorname{Re}\phi(x)$ is small. Note that $\langle S_\varepsilon x,x\rangle$ can be small even $|x|$ is big. Note also that

$\{\langle S_\varepsilon x,x\rangle < a\}$ is an ellipsoid with semi-axes $(a/\lambda_1,)^{\frac{1}{2}}, (a/\lambda_2)^{\frac{1}{2}}, \cdots$, where λ_n's are eigenvalues of S_ε.

Proof. __Necessity__. Let $\phi(x) = \int_H e^{i\langle x,y\rangle}\,\mu(dy)$.

Exercise 9 shows (a). Let $\varepsilon > 0$ be given. Choose $0 < r < \infty$.

such that $\mu(S_r) > 1-\varepsilon/2$, where S_r is the ball $\{\ |x| \leq r\}$.

Then $\phi(x) = \int_{S_r} e^{i\langle x,y\rangle}\,\mu(dy) + \int_{S_r^c} e^{i\langle x,y\rangle}\,\mu(dy)$.

Note that $\left|\int_{S_r^c} e^{i\langle x,y\rangle}\,\mu(dy)\right| \leq \mu(S_r^c) < \varepsilon/2$. Hence it is sufficient to show that

$$1 - \mathrm{Re}\int_{S_r} e^{i\langle x,y\rangle}\,\mu(dy) \leq \langle S_\varepsilon x,x\rangle + \varepsilon/2$$

for some operator $S_\varepsilon \in \mathcal{S}$.

But $$1 - \mathrm{Re}\int_{S_r} e^{i\langle x,y\rangle}\,\mu(dy) = \int_{S_r} (1 - \cos\langle x,y\rangle)\,\mu(dy) + \mu(S_r^c)$$

$$\leq \int_{S_r} (1 - \cos\langle x,y\rangle)\,\mu(dy) + \varepsilon/2.$$

Recall that $1 - \cos\theta \leq \frac{1}{2}\theta^2$ for all real θ. Hence

$$\int_{S_r} (1 - \cos\langle x,y\rangle)\,\mu(dy) \leq \frac{1}{2}\int_{S_r} \langle x,y\rangle^2\,\mu(dy).$$

The same argument in the proof of Theorem 2.1 shows that there exists $S_\varepsilon \in \mathcal{S}$ such that

$$\langle S_\varepsilon x,y\rangle = \frac{1}{2}\int_{S_r} \langle x,z\rangle\langle y,z\rangle\,\mu(dz).$$

The desired conclusion follows immediately.

Remarks. (1) Observe that if the covariance operator S_μ of μ exists, then we have $1 - \mathrm{Re}\,\phi(x) \leq \frac{1}{2}\langle S_\mu x,x\rangle$. But S_μ, in general, is not an S-operator. Of course, if $S_\mu \in \mathcal{S}$ then (b) is trivially satisfied by taking $S_\varepsilon = \frac{1}{2}S_\mu$ for all $\varepsilon > 0$.

(2) Observe also that S_ε is the covariance operator of the following Borel measure ν in H,

$$\nu(E) = \frac{1}{2}\mu\ (E \cap S_r), \quad E \in \mathcal{B}(H).$$

Clearly, ν satisfies the hypothesis of Theorem 2.1 and thus its covariance operator S_ε is an S-operator. This is another proof for $S_\varepsilon \in \mathcal{S}$.

Sufficiency. (This is a little bit involved, and we have to use Bochner's theorem for the finite dimensional space $\mathbb{R}^n$.)

Step 1. We first derive some properties implied by (a).

(a-1) $|\phi(x)| \leq 1$, and $\phi(x) = \overline{\phi(-x)}$ for all x in H,

(a-2) $|\phi(x) - \phi(y)| \leq 2\ \sqrt{|1 - \phi(x-y)|}$ for all x,y in H,

(a-3) $|1 - \phi(x)| \leq \sqrt{2}\ \sqrt{1 - \operatorname{Re}\ \phi(x)}$ for all x in H.

(a-1) : Take $n = 2$, $x_1 = 0$ and $x_2 = x$. (a) implies that the following matrix is positive definite in the sense of linear algebra,

$$\begin{pmatrix} 1 & \phi(x) \\ \phi(-x) & 1 \end{pmatrix}.$$

Hence $\phi(x) = \overline{\phi(-x)}$. Moreover, the determinant is non-negative, i.e. $1 - \phi(x)\ \overline{\phi(x)} \geq 0$. Hence $|\phi(x)| \leq 1$.

(a-2) : Take $n = 3$ and $x_1 = 0$, $x_2 = x$ and $x_3 = y$. (a) implies that the matrix

$$\begin{pmatrix} 1 & \phi(x) & \phi(y) \\ \phi(-x) & 1 & \phi(y-x) \\ \phi(-y) & \phi(x-y) & 1 \end{pmatrix} = \begin{pmatrix} 1 & \phi(x) & \phi(y) \\ \overline{\phi(x)} & 1 & \overline{\phi(x-y)} \\ \overline{\phi(y)} & \phi(x-y) & 1 \end{pmatrix}$$

is positive definite. Hence the determinant $D \geq 0$. But

$$D = 1 + \phi(x)\,\overline{\phi(y)}\,\overline{\phi(x-y)} + \overline{\phi(x)}\phi(y)\phi(x-y) - |\phi(y)|^2 - |\phi(x)|^2 - |\phi(x-y)|^2$$

$$= 1 + 2\,\mathrm{Re}\left[\phi(x)\overline{\phi(y)}\,\overline{\phi(x-y)}\right] - (|\phi(x)|^2 + |\phi(y)|^2) - |\phi(x-y)|^2$$

$$= 1 + 2\,\mathrm{Re}\left[\phi(x)\overline{\phi(y)}\,\overline{\phi(x-y)}\right] - \left(|\phi(x) - \phi(y)|^2 + 2\,\mathrm{Re}\,\phi(x)\overline{\phi(y)}\right) - |\phi(x-y)|^2$$

$$= 1 + 2\,\mathrm{Re}\left[\phi(x)\overline{\phi(y)}\,(\overline{\phi(x-y)} - 1)\right] - |\phi(x) - \phi(y)|^2 - |\phi(x-y)|^2.$$

Note that

$$1 - |\phi(x-y)|^2 = (1 + |\phi(x-y)|)(1 - |\phi(x-y)|)$$

$$\leq (1+1)(|\,1 - \phi(x-y)|)$$

$$= 2|\,1 - \phi(x-y)|,$$

and $2\,\mathrm{Re}\left[\phi(x)\overline{\phi(y)}(\overline{\phi(x-y)} - 1)\right] \leq 2\,|\phi(x)|\,|\overline{\phi(y)}|\,|\overline{\phi(x-y)} - 1| \leq 2|1 - \phi(x-y)|.$

Hence $0 \leq D \leq 4\,|1 - \phi(x-y)| - |\phi(x) - \phi(y)|^2$,

i.e. $|\phi(x) - \phi(y)| \leq 2\sqrt{|\,1 - \phi(x-y)\,|}.$

(a-3) : Note that if $|\,z\,| \leq 1$ then

$$|\,1 - z|^2 = (1-z)(1-\overline{z}) = 1 - (z + \overline{z}) + |\,z\,|^2$$

$$\leq 1 - 2\,\mathrm{Re}\,z + 1 = 2(1 - \mathrm{Re}\,z).$$

From (a-1) we know that $|\,\phi(x)\,| \leq 1$, therefore,

$$|\,1 - \phi(x)|^2 \leq 2\,(1 - \mathrm{Re}\,\phi(x)).$$

Hence, $|1 - \phi(x)| \leq \sqrt{2}\ \sqrt{1 - \mathrm{Re}\ \phi(x)}$.

Remark . (a-2) says that if ϕ is continuous at the origin (w.r.t. whatever topology) then ϕ is continuous in the whole H (w.r.t. the same topology). (a-3) says that the continuity of Re ϕ implies the continuity of ϕ . Compare (b) of this theorem.

Step 2. Let $\{e_n\}$ be a fixed orthonormal basis of H. For each $n \geq 1$, Define $\psi_{e_1,\ldots,e_n}$ in $\mathbb{R}^n$ by

$$\psi_{e_1,\ldots,e_n}(a_1,\ldots,a_n) = \phi(a_1 e_1 + \cdots + a_n e_n).$$

Note that Re ϕ is continuous at the origin by (b). Hence ϕ is continuous in H by (a-2) and (a-3). Hence $\psi_{e_1,\ldots e_n}$ is continuous for each n. Also, $\psi_{e_1,\ldots,e_n}$ is positive definite and $\psi_{e_1,\ldots,e_n}(0) = 1$. Bochner's theorem for $\mathbb{R}^n$ gives us a family of probability measures $\{\mu_n\}$ such that

$$\psi_{e_1,\ldots,e_n}(a_1,\ldots,a_n) = \int_{\mathbb{R}^n} e^{i(a,y)}\mu_n(dy),\ a = (a_1,\ldots,a_n).$$

It is easy to see that $\{\mu_n\}$ is a consistent family. Hence Kolmogorov's theorem implies the existence of a probability space (Ω, P) and a sequence of random variables $\{X_n\}$ such that

$$\mu_n = P\circ(X_1,\ldots,X_n)^{-1},\ n = 1,2,\ldots .$$

Therefore,

$$\phi(a_1e_1 + a_2e_2 + \cdots + a_ne_n) = \int_{\mathbb{R}^n} e^{i(a,y)}\ \mu_n(dy),\ a = (a_1,a_2,\cdots,a_n)$$

$$= \int_\Omega e^{i(a_1X_1 + a_2X_2 + \cdots + a_nX_n)} dP.$$

Step 3. Suppose we can show that $\sum_{n=1}^{\infty} X_n^2 < \infty$ almost surely (this will be shown in Step 4) then we are done. To see this, define

$$X(\omega) = \sum_{n=1}^{\infty} X_n(\omega)e_n, \quad \omega \in \Omega .$$

Then X is measurable from Ω into H. Define $\mu = P\circ X^{-1}$. μ is a probability Borel measure of H. Let Q_n be the orthogonal projection of H into the span of $e_1,e_2,\ldots,e_n$; i.e.

$$Q_n x = \langle x,e_1\rangle\ e_1 + \langle x,e_2\rangle\ e_2 + \cdots + \langle x,e_n\rangle\ e_n,\quad x \in H.$$

Then

$Q_nX = \sum_{k=1}^{n} X_k e_k$. By step 2, we have

$$\phi(Q_n x) = \int_\Omega e^{i\langle x,Q_n X\rangle} dP.$$

Now, letting $n \to \infty$ and observing $Q_nx \to x$ in H as $n \to \infty$, we have $\phi(Q_nx) \to \phi(x)$ by the continuity of ϕ. Apply Lebesgue's dominated convergence theorem,

$$\int_\Omega e^{i\langle x,Q_n X\rangle} dP \longrightarrow \int_\Omega e^{i\langle x,X\rangle} dP \text{ as } n \to \infty.$$

Hence we have

$$\phi(x) = \int_\Omega e^{i\langle x,X\rangle} dP = \int_\Omega e^{i\langle x,y\rangle} \mu(dy),\ x \in H.$$

Step 4. To show that $\sum_{n=1}^{\infty} X_n^2 < \infty$ almost surely.

First note that

$$\int_{\mathbb{R}^n} e^{i(a_1y_1 + \cdots + a_ny_n)} \left(\frac{1}{\sqrt{2\pi}}\right)^n e^{-\frac{1}{2}\{y_1^2 + \cdots + y_n^2\}} dy_1 \cdots dy_n$$

$$= e^{-\frac{1}{2}\{a_1^2 + \cdots + a_n^2\}}.$$

Therefore,

$$\int_\Omega e^{-\frac{1}{2}\{X_{k+1}^2 + \cdots + X_{k+n}^2\}} dP$$

$$= \int_\Omega \left[\int_{\mathbb{R}^n} e^{i\sum_{j=1}^n X_{k+j}y_j} \left(\frac{1}{\sqrt{2\pi}}\right)^n e^{-\frac{1}{2}\sum_{j=1}^n y_j^2} dy_1 \cdots dy_n\right] dP$$

$$= \int_{\mathbb{R}^n} \left[\int_\Omega e^{i\sum_{j=1}^n X_{k+j}y_j} dP\right] \left(\frac{1}{\sqrt{2\pi}}\right)^n e^{-\frac{1}{2}\sum_{j=1}^n y_j^2} dy_1 \cdots dy_n$$

$$= \int_{\mathbb{R}^n} \phi(y_1e_{k+1} + \cdots + y_ne_{k+n}) \left(\frac{1}{\sqrt{2\pi}}\right)^n e^{-\frac{1}{2}\{y_1^2 + \cdots + y_n^2\}} dy_1 \cdots dy_n$$

$$= \int_{\mathbb{R}^n} \mathrm{Re}\, \phi(y_1e_{k+1} + \cdots + y_ne_{k+n}) \left(\frac{1}{\sqrt{2\pi}}\right)^n e^{-\frac{1}{2}\{y_1^2 + \cdots + y_n^2\}} dy_1 \cdots dy_n,$$

since the left hand side is real.

Now, let $\varepsilon > 0$ be given, we use assumption (b) to get $S_\varepsilon \in \mathcal{S}$ such that $1 - \mathrm{Re}\, \phi(x) \leq \langle S_\varepsilon x, x\rangle + \varepsilon$, $x \in H$. Therefore,

$$1 - \int_\Omega e^{-\frac{1}{2}\{X_{k+1}^2 + \cdots + X_{k+n}^2\}} dP$$

$$\leq \int_{\mathbb{R}^n} (1 - \mathrm{Re}\, \phi(y_1e_{k+1} + \cdots + y_ne_{k+n})) p_1(dy),$$

where $p_1(dy) = (\frac{1}{\sqrt{2\pi}})^n e^{-\frac{1}{2}\{y_1^2 + \cdots + y_n^2\}} dy_1 \ldots dy_n$ is a probability measure in $\mathbb{R}^n$.

The above

$$\leq \int_{\mathbb{R}^n} [< S_\varepsilon(y_1 e_{k+1} + \cdots + y_n e_{k+n}), y_1 e_{k+1} + \cdots + y_n e_{k+n} > + \varepsilon] p_1(dy)$$

$$= \varepsilon + \int_{\mathbb{R}^n} \sum_{i,j=1}^{n} y_i y_j < S_\varepsilon e_{k+i}, e_{k+j} > p_1(dy)$$

$$= \varepsilon + \sum_{i,j=1}^{n} < S_\varepsilon e_{k+i}, e_{k+j} > \int_{\mathbb{R}^n} y_i y_j \, p_1(dy)$$

$$= \varepsilon + \sum_{j=1}^{n} < S_\varepsilon e_{k+j}, e_{k+j} > .$$

Let $n \to \infty$, apply Lebesgue's dominated convergence theorem to the left hand side and remember that $S_\varepsilon \in \mathcal{S}$. Hence

$$1 - \int_\Omega e^{-\frac{1}{2}\sum_{j=1}^{\infty} X_{k+j}^2} dP \leq \varepsilon + \sum_{m=k+1}^{\infty} < S_\varepsilon e_m, e_m >$$

$$\leq 2\varepsilon \quad \text{whenever } k \geq k_0, \text{ say.}$$

Hence

$$\int_\Omega e^{-\frac{1}{2}\sum_{j=1}^{\infty} X_{k+j}^2} dP \geq 1 - 2\varepsilon, \; k \geq k_0.$$

Finally,

$$P\{\sum_{n=1}^{\infty} X_n^2 < \infty\} \geq \int_{\sum_{n=1}^{\infty} X_n^2 < \infty} e^{-\frac{1}{2}\sum_{j=1}^{\infty} X_{k+j}^2} dP$$

$$= \int_\Omega e^{-\frac{1}{2}\sum_{j=1}^{\infty} X_{k+j}^2} dP \quad \text{(why ?)}$$

$$\geq 1 - 2\varepsilon.$$

Thus $P\{X_1^2 + \cdots + X_n^2 + \cdots < \infty\} \geq 1 - 2\varepsilon$ for any $\varepsilon > 0$.

Of course, we must have

$$P\{\sum_{n=1}^{\infty} X_n^2 < \infty\} = 1,$$

which is what we wanted to prove. #

Now, we are going to study Gaussian measures in H. We start with the following

<u>Definition 2.6</u>. A <u>Gaussian measure</u> μ in H is a Borel measure in H such that for each $x \in H$, the measurable function $\langle x, \cdot \rangle$ is normally distributed, i.e. there exist real numbers m_x and σ_x such that

$$\mu\{y \in H;\ \langle x,y\rangle \le a\} = \int_{-\infty}^{a} \frac{1}{\sqrt{2\pi\sigma_x}} e^{-\frac{(t-m_x)^2}{2\sigma_x}} dt.$$

<u>Lemma 2.1</u>. Let μ be a Gaussian measure in H. Then its characteristic functional is given by

$$\phi(x) = e^{i\langle m_\mu, x\rangle - \frac{\langle S_\mu x, x\rangle}{2}}.$$

where m_μ is the mean of μ and S_μ the covariance operator of μ.

Remarks. (1) Definition 2.6 does not say anything about the existence of m_μ and S_μ. The lemma says, of course, if m_μ and S_μ exist then ϕ must be given by the above form.

(2) It can be shown that $\hat{\mu} = \hat{\nu} \Rightarrow \mu = \nu$ for any two Borel measures in H. Then, in view of the above lemma, a Gaussian measure in H is uniquely determined by its mean and

its covariance operator.

Proof. $\phi(x) = \int_H e^{i\langle x,y\rangle} \mu(dy) = \int_{-\infty}^{\infty} e^{it} \mu_x(dt)$,

where μ_x is the distribution of $\langle x,\cdot\rangle$.

Hence,

$$\phi(x) = \int_{-\infty}^{\infty} e^{it} \frac{1}{\sqrt{2\pi\sigma_x}} e^{-\frac{(t-m_x)^2}{2\sigma_x}} dt.$$

Make a change of variables first and use contour integration to conclude

$$\phi(x) = e^{i m_x - \frac{\sigma_x}{2}}.$$

But

$$m_x = \int_{-\infty}^{\infty} t\, \mu_x(dt) = \int_H \langle x,y\rangle \mu(dy)$$

$$= \langle x, m_\mu\rangle = \langle m_\mu, x\rangle,$$

and

$$\sigma_x = \int_{-\infty}^{\infty} t^2 \mu_x(dt) = \int_H \langle x,y\rangle^2 \mu(dy)$$

$$= \langle S_\mu x, x\rangle.$$

Hence

$$\phi(x) = e^{i\langle m_\mu, x\rangle - \frac{\langle S_\mu x,x\rangle}{2}}. \qquad \#$$

Theorem 2.3. (Prohorov) (a) If μ is a Gaussian measure in H then $S_\mu \in \mathcal{S}$. (b) If $x_o \in H$ and $S \in \mathcal{S}$ then $\phi(x) = e^{i\langle x_o,x\rangle - \frac{1}{2}\langle Sx,x\rangle}$ is the characteristic functional of a (Gaussian) measure in H.

Remark. (i) It follows from Theorem 2.1 and (a) that $\int_H |x|^2 \mu(dx) < \infty$. Therefore,

$\int_H |x| \mu(dx) \le (\int_H |x|^2 \mu(dx))^{1/2} < \infty$ and by the Remark following Definition 2.3, m_μ exists.

(ii) The theorem can be rephrased as follows :

$\phi(x) = e^{i<x_0,x> - \frac{1}{2}<Sx,x>}$ is the characteristic functional of a Gaussian measure iff $S \in \mathcal{S}$.

Proof. (a) By the proof of Lemma 2.1 we have

$$\phi(x) = e^{i\, m_x - \frac{1}{2}\sigma_x}.$$

Let $0 < \varepsilon < \frac{1}{2}$ be given. By Theorem 2.2, there exist $S_\varepsilon \in \mathcal{S}$ such that

$$1 - \mathrm{Re}\ \phi(x) \leq <S_\varepsilon x,x> + \varepsilon \text{ for all } x \in H.$$

Let $\lambda_n, n = 1,2, \cdots$ be non-zero eigenvalues of S_ε and $\{e_n\}$ the corresponding unit eigenvectors. Let $\{f_n\}$ be an orthonormal basis of ker S_ε. Define

$$S\,x = \Sigma\,\lambda_n <x,e_n>\,e_n + \Sigma\,\frac{1}{n^2}<x,f_n>\,f_n.$$

Then $S \in \mathcal{S}$, ker $S = \{0\}$ and $<S_\varepsilon x,x> \leq <Sx,x>$ for all $x \in H$.

Observe also that

$$1 - e^{-\frac{1}{2}\sigma_x} \leq 1 - \mathrm{Re}\ \phi(x).$$

Hence

$$1 - e^{-\frac{1}{2}\sigma_x} \leq <Sx,x> + \varepsilon \text{ for all } x \text{ in } H.$$

<u>Claim</u> : $\sigma_x \leq (\frac{4}{\varepsilon}\log\frac{1}{1-2\varepsilon}) <S\,x,x>$ for all x in H.

Suppose $<S\,x,x> < \varepsilon$. Then

$$1 - e^{-\frac{1}{2}\sigma_x} \leq 2\varepsilon,$$

or
$$-\frac{1}{2}\sigma_x \geq \log (1 - 2\varepsilon).$$

Hence
$$\sigma_x \leq 2 \log \frac{1}{1-2\varepsilon}, \text{ whenever } \langle S x,x \rangle < \varepsilon.$$

Let $x \neq 0$ be arbitrary. Consider $y = (\frac{\varepsilon}{2\langle S x,x\rangle})^{\frac{1}{2}} x$.

Then $\langle Sy,y \rangle < \varepsilon$. Therefore,

$$\sigma_y \leq 2 \log \frac{1}{1-2\varepsilon}.$$

But
$$\sigma_y = \int_H \langle y,z \rangle^2 \mu(dz)$$

$$= \frac{\varepsilon}{2\langle S x,x\rangle} \int_H \langle x,z \rangle^2 \mu(dz)$$

$$= \frac{\varepsilon}{2\langle S x,x \rangle} \sigma_x.$$

So,
$$\sigma_x \leq (\frac{4}{\varepsilon} \log \frac{1}{1-2\varepsilon}) \langle Sx,x\rangle.$$

This gives the above claim.

Let $c = \frac{4}{\varepsilon} \log \frac{1}{1-2\varepsilon}$. Then the above claim says that

$$\int_H \langle x,y \rangle^2 \mu(dy) \leq c \langle S x,x \rangle.$$

This inequality shows not only the existence of the covariance operator S_μ of μ, but also $S_\mu \in \mathcal{S}$.

(b) Clearly, $\phi(0) = 1$ and ϕ is positive definite.

Consider first the case $x_o = 0$. Then

$$1 - \operatorname{Re} \phi(x) = 1 - e^{-\frac{\langle Sx,x\rangle}{2}} \leq \frac{1}{2}\langle Sx,x\rangle$$

because $1 - e^{-y} \leq y$ for all $y \geq 0$.

Since $\frac{1}{2} S \in \mathcal{S}$, (b) of Theorem 2.2 is trivially satisfied. Therefore by the conclusion of Theorem 2.2, there exists a Borel measure μ such that $\phi(x) = \hat{\mu}(x)$, $x \in H$. That is,

$$\int_H e^{i\langle x,y\rangle} \mu(dy) = e^{-\frac{\langle Sx,x\rangle}{2}}, \quad x \in H.$$

Let μ_x be the distribution of $\langle x,\cdot\rangle$. Then

$$\int_{-\infty}^{\infty} e^{it} \mu_x(dt) = e^{-\frac{\langle Sx,x\rangle}{2}}.$$

Hence μ_x is a normal distribution with mean 0 and variance $\langle Sx,x\rangle$. That is, μ is a Gaussian measure in H. Consider now the case $x_o \neq 0$. Let $\psi(x) = e^{-\frac{1}{2}\langle Sx,x\rangle}$. Then

$$\phi(x) = e^{i\langle x_o,x\rangle} \psi(x).$$

By the case already proved, there exists a Gaussian measure ν in H such that $\psi(x) = \hat{\nu}(x)$, $x \in H$. Define a Borel measure μ in H as follows :

$$\mu(E) = \nu(E - x_o), \quad E \in \mathcal{B}(H).$$

It is easy to see that μ is a Gaussian measure and

$$\hat{\mu}(x) = e^{i\langle x_o,x\rangle} \hat{\nu}(x) = e^{i\langle x_o,x\rangle} \psi(x) = \phi(x). \quad \#$$

<u>Corollary 2.1</u>. Suppose $\dim H = \infty$ then there is no Borel measure μ in H such that $\hat{\mu}(x) = e^{-\frac{|x|^2}{2}}$.

Proof: If there is such a μ then $\hat{\mu}(x)$ must satisfy Theorem 2.2 (b) and the same trick as in the proof (a) of Theorem 2.3 shows that $I \in \mathcal{S}$. But we know that when $\dim H = \infty$, the identity operator I is not in $\mathcal{S}$.

Now we show the invariance property of Gaussian measures we mentioned in the introduction. Let μ be a Gaussian measure in H with mean 0. We may assume its covariance operator S is injective by considering the support of μ, if necessary. For the definition of support of μ and its properties, see [23].

In the image $\sqrt{S}(H)$, we define an inner product $\langle\ ,\ \rangle_o$ by

$$\langle \sqrt{S}x, \sqrt{S}y \rangle_o = \langle x,y \rangle . \quad x,y \in H.$$

Note that $\sqrt{S}$ is strictly positive definite, and is a Hilbert-Schmidt operator. $\sqrt{S}(H)$ is a Hilbert space with inner product $\langle \cdot,\cdot \rangle_o$. In fact, $\sqrt{S}$ is an isometry from H onto $\sqrt{S}(H)$. Hence $\sqrt{S}$ is unitary, as an operator from H to $\sqrt{S}(H)$.

Notation. (1) $H_o = \sqrt{S}(H)$. Later on, we will regard (H,μ) as a couple (H_o,H).

(2) Let $U \in \mathcal{L}(H)$. We use $\tilde{U}$ to denote the adjoint of U. If H_o is invariant under U, i.e. $U(H_o) \subset H_o$, then $U|_{H_o}$ can be considered as an operator of H_o. An easy application of the closed graph theorem shows that $U|_{H_o} \in \mathcal{L}(H_o)$. This operator will be denoted by U_o. The adjoint of $V \in \mathcal{L}(H_o)$ will be denoted by V^*.

Lemma 2.2. Let $U \in \mathcal{L}(H)$ and $U(H_o) \subset H_o$. Then $U_o^* = S\tilde{U}S^{-1}$.

Proof. For $x,y \in H_o$,

$$\langle x, U_o^* y\rangle_o = \langle U_o x, y\rangle_o = \langle(\sqrt{S})^{-1} U_o x, (\sqrt{S})^{-1} y\rangle = \langle U_o x, S^{-1} y\rangle$$

$$= \langle Ux, S^{-1} y\rangle = \langle x, \tilde{U} S^{-1} y\rangle = \langle\sqrt{S}x, \sqrt{S}\, \tilde{U} S^{-1} y\rangle_o$$

$$= \langle x, S\tilde{U} S^{-1} y\rangle_o .$$ #

Theorem 2.4. Let $U \in \mathcal{L}(H)$ and $U(H_o) \subset H_o$. Suppose U_o is unitary in H_o. Then $\mu U^{-1} = \mu$.

Proof. It is easy to see that μU^{-1} is a Gaussian measure in H with mean 0. Therefore, all we have to show is that μU^{-1} has the same covariance operator S as μ. Let T be the covariance operator of μU^{-1}. Then for $x,y \in H$,

$$\langle Tx,y\rangle = \int_H \langle x,z\rangle \langle y,z\rangle \mu U^{-1}(dz)$$

$$= \int_H \langle x,Uz\rangle \langle y,Uz\rangle \mu(dz)$$

$$= \int_H \langle \tilde{U}x,z\rangle \langle \tilde{U}y,z\rangle \mu(dz)$$

$$= \langle S \tilde{U} x, \tilde{U}y\rangle$$

$$= \langle U S \tilde{U} x,y\rangle = \langle U_o S \tilde{U} x,y\rangle .$$

But since U_o is H_o-unitary, we have $U_o U_o{}^* = I_{H_o}$.

Hence, by Lemma 2.2,

$$U_o S \tilde{U} S^{-1} = I_{H_o} .$$

Therefore, we have $U_o S \tilde{U} = S$. Hence $\langle T x, y\rangle = \langle S x, y\rangle$ for all x, y in H, and $T = S$. #

Remarks. (1) Let $S x = \sum_{n=1}^{\infty} \alpha_n \langle x, e_n\rangle e_n$. $\alpha_n > 0$ for all n because S is injective. In this case, the characteristic functional ϕ of μ is given by

$$\phi(x) = e^{-\frac{1}{2}\langle Sx, x\rangle} = e^{-\frac{1}{2}\sum_{n=1}^{\infty}\alpha_n \langle x, e_n\rangle^2}.$$

(2) If S is given as above, then

$$H_o = \{x = \sum_{n=1}^{\infty} \beta_n e_n ;\ \sum_{n=1}^{\infty} \frac{\beta_n^2}{\alpha_n} < \infty\}.$$

§3. Wiener measure and Wiener integral in $C[0,1]$.

Let $C[0,1]$ denote the set of real-valued functions $x(t)$ in the unit interval $[0,1]$ with $x(0) = 0$. $C[0,1]$ is a Banach space with the supremum norm $\|x\| = \sup_{0\le t\le 1} |x(t)|$. Let $\mathcal{B}$ denote the Borel field of $C[0,1]$. A subset I of $C[0,1]$ of the following form

$$I = \{ x \in C[0,1];\ \left(x(t_1), x(t_2), \ldots, x(t_n)\right) \in E\},$$

where $0 < t_1 < t_2 < \cdots < t_n \le 1$ and E is a Borel subset of $\mathbb{R}^n$, will be called a cylinder set. Obviously, the collection $\mathcal{R}$ of cylinder subsets of $C[0,1]$ is a field, but not a σ-field. As a matter of fact, we will see later on that the σ-field generated by $\mathcal{R}$ is the Borel field $\mathcal{B}$.

Definition 3.1. Let I be the cylinder set given above. Define

$$w(I) = \left[(2\pi)^n t_1 (t_2 - t_1) \cdots (t_n - t_{n-1})\right]^{-\frac{1}{2}} \int_E \exp\left\{ - \left[\frac{u_1^2}{t_1} + \frac{(u_2-u_1)^2}{t_2-t_1} + \cdots + \frac{(u_n-u_{n-1})^2}{t_n-t_{n-1}} \right]/2 \right\} du_1\, du_2 \ldots du_n.$$

Remark. Obviously, w is finitely additive in the field $\mathcal{R}$. The main theorem in this section is about the σ-additivity of w. Thus, w has a unique extension, still denoted by w, to the Borel field $\mathcal{B}$.

Definition 3.2. w is called the Wiener measure in $C[0,1]$. The integral in $C[0,1]$ with respect to w is called Wiener integral.

If f is a Wiener integrable function, its integral will be denoted by $E_w[f] \equiv \int_{C[0,1]} f(x)\, w(dx)$.

Before proving the σ-additivity of w, we give some examples of Wiener integral so that we will have a more concrete feeling about Wiener measure. Note that if $0 < t \leq 1$ then

$$w(\{x \in C[0,1];\ a \leq x(t) \leq b\}) = \frac{1}{\sqrt{2\pi t}} \int_a^b e^{-\frac{x^2}{2t}}\, dx,$$

so the functional $f(x) = x(t)$ is normally distributed with mean 0 and variance t.

Moreover, let $0 < s < t \leq 1$ be fixed and consider the random variable $x(t) - x(s)$. Let $E \subset \mathbb{R}^2$ be the set $E = \{(x,y);\ a \leq x-y \leq b\}$. Then from the definition of w,

$$w(\{x;\ a \leq x(t) - x(s) \leq b\}) = x(\{x; (x(t),x(s)) \in E\})$$

$$= \frac{1}{\sqrt{(2\pi)^2\, s(t-s)}} \iint_E e^{-\frac{1}{2}\left\{\frac{v^2}{s} + \frac{(u-v)^2}{t-s}\right\}}\, du\, dv$$

$$= \frac{1}{\sqrt{(2\pi)^2 s(t-s)}} \int_{-\infty}^{\infty} \int_{v+a}^{v+b} e^{-\frac{1}{2}\left\{\frac{v^2}{s} + \frac{(u-v)^2}{t-s}\right\}}\, du\, dv.$$

Make a change of variables, $u - v = \tau_1$ and $u = \tau_2$, to obtain

$$w\{a \leq x(t)-x(s) \leq b\} = \frac{1}{\sqrt{(2\pi)^2 s(t-s)}} \int_{-\infty}^{\infty} \int_a^b e^{-\frac{1}{2}\left\{\frac{\tau_2^2}{s} + \frac{\tau_1^2}{t-s}\right\}}\, d\tau_1 d\tau_2.$$

$$= \frac{1}{\sqrt{2\pi(t-s)}} \int_a^b e^{-\frac{\tau_1^2}{2(t-s)}}\, d\tau_1 \cdot \frac{1}{\sqrt{2\pi s}} \int_{-\infty}^{\infty} e^{-\frac{\tau_2^2}{2s}}\, d\tau_2$$

$$= \frac{1}{\sqrt{2\pi(t-s)}} \int_a^b e^{-\frac{\tau^2}{2(t-s)}} d\tau.$$

Therefore, we see that the random variable $x(t) - x(s)$ is normally distributed with mean 0 and variance $t-s$, $(t > s)$. We have shown the following

Example 1. $\int_{C[0,1]} (x(t)-x(s))w(dx) = 0, \int_{C[0,1]} (x(t)-x(s))^2 w(dx) = |t-s|.$

Exercise 10. Show that if $0 \leq t \leq s \leq v \leq u \leq 1$ then the random variables $x(s) - x(t)$ and $x(u) - x(v)$ are independent.

Example 2.

$$\int_{C[0,1]} x(t)x(s)w(dx) = \min(t,s).$$

Proof . Assume $t \leq s$, then

$$x(t)x(s) = x(t)(x(s) - x(t)) + x(t)^2.$$

Hence

$$E_w[x(t)x(s)] = E_w[x(t)(x(s) - x(t))] + E_w[x(t)^2].$$

But $E_w[x(t)(x(s)-x(t))] = E_w[(x(t)-x(0))(x(s)-x(t))]$

$= E_w[x(t)-x(0)] \times E_w[x(s)-x(t)]$ (by Exercise 10)

$= 0.$ (by Example 1)

Also, $E_w[x(t)^2] = t$ by Example 1 .

Hence $E_w[x(t)x(s)] = t = \min(t,s)$. #

Exercise 11. Let $0 \leq s < t \leq 1$ then

$$\int_{C[0,1]} \{x(t)-x(s)\}^p \, w(dx) = 0 \text{ if } p \text{ is an odd natural number}$$

and $$\int_{C[0,1]} |x(t)-x(s)|^p \, w(dx) = \frac{1}{\sqrt{\pi}} \sqrt{2^p (t-s)^p} \; \Gamma\left(\frac{p}{2} + \frac{1}{2}\right),$$

where Γ is the gamma function.

Example 3. $$\int_{C[0,1]} \left[\int_0^1 x(t)\,dt\right] w(dx) = 0$$

and $$\int_{C[0,1]} \left[\int_0^1 x(t)^2 \, dt\right] w(dx) = \frac{1}{2}$$

Proof. Use Fubini theorem and Example 1. #

Some more complicated examples will be given later on. We want to show now that w has a σ-additive extension to the σ- field generated by the cylinder sets.

Notation 1. $C = C[0,1]$

2. S = binary rationals in $[0,1]$.

3. $C_\alpha = \{ x \in C; \exists\, a=a(x) \text{ s.t. } |x(t)-x(s)| \leq a|t-s|^\alpha \; \forall t,s \}$.

4. $B_\alpha = \{ x \in C; \exists\, a=a(x) \text{ s.t. } |x(t)-x(s)| \leq a|t-s|^\alpha \; \forall t,s \in S \}$.

5. $H_\alpha[a] = \{ x \in C; \exists\, s_1,s_2 \in S \ni |x(s_1)-x(s_2)| > a|s_1-s_2|^\alpha \}$.

6. $H_\alpha = \{ x \in C; \forall a > 0, \exists\, s_1,s_2 \in S \ni |x(s_1)-x(s_2)| > a|s_1-s_2|^\alpha \}$.

7. $I_{\alpha,a,k,n} = \{ x \in C; |x(\frac{k}{2^n}) - x(\frac{k-1}{2^n})| > a(\frac{1}{2^n})^\alpha \}$,

$k = 1,2,3,\ldots 2^n$.

8. w^* = the outer measure of w.

The following lemma is obvious from the above definitions.

Lemma 3.1. (a) $0<\alpha<\beta \Rightarrow C_\beta \subset C_\alpha \subset C$,

(b) $C_\alpha = B_\alpha$, $\alpha>0$,

(c) $H_\alpha = \bigcap_{a>0} H_\alpha[a] = \bigcap_{n=1}^{\infty} H_\alpha[a_n]$, $a_n > 0$, $a_n \nearrow \infty$,

(d) $H_\alpha = C \setminus B_\alpha$.

Lemma 3.2. Let $\alpha>0$ and $a>0$. If $x \in C[0,1]$ satisfies

$$\left|x\left(\frac{k}{2^n}\right) - x\left(\frac{k-1}{2^n}\right)\right| \le a\left(\frac{1}{2^n}\right)^\alpha \quad \forall k=0,1,\ldots 2^n \text{ and } \forall n=1,2,\ldots.$$

then

$$|x(s_1) - x(s_2)| \le 2a \frac{1}{1-2^{-\alpha}} |s_1 - s_2|^\alpha \quad \forall s_1, s_2 \in \mathcal{S}.$$

Proof. If $s_1 = 0$ and $s_2 = 1$ we have nothing to prove since $1-2^{-\alpha} \le 2$. Hence assume that $s_1 < s_2$ and $[s_1,s_2] \neq [0,1]$. Note that every $s \in \mathcal{S}$ is expressed uniquely as $\frac{k}{2^n}$ for k odd. It is easy to see that there is a unique $s_o \in \mathcal{S}$ with $s_1 \le s_o \le s_2$ and $s_o = \frac{q}{2^p}$ (q : odd) has smallest p.

Now, if $s_o \neq s_1$, then

$$s_o - s_1 = \frac{1}{2^{m_1}} + \frac{1}{2^{m_2}} + \ldots + \frac{1}{2^{m_j}}, \quad m_1 < m_2 < \ldots < m_j$$

and if $s_o \neq s_2$, then

$$s_2 - s_o = \frac{1}{2^{n_1}} + \frac{1}{2^{n_2}} + \ldots + \frac{1}{2^{n_k}}, \quad n_1 < n_2 < \ldots < n_k.$$

Consider the following intervals,

$$[s_1, s_1 + \frac{1}{2^{m_j}}], [s_1 + \frac{1}{2^{m_j}}, s_1 + \frac{1}{2^{m_{j-1}}} + \frac{1}{2^{m_j}}], \ldots\ldots\ldots\ldots$$

$$\ldots\ldots [s_o - \frac{1}{2^{m_1}}, s_o]$$

and $[s_o, s_o + \frac{1}{2^{n_1}}], \ldots, [s_o + \frac{1}{2^{n_1}} + \cdots + \frac{1}{2^{n_{k-1}}}, s_2]$.

Let $p = \min(m_1, n_1)$ and $q = \max(m_j, n_k)$. Then it is easy to see that

$$|x(s_1) - x(s_2)| \leq 2a \sum_{k=p}^{q} \left(\frac{1}{2^k}\right)^{\alpha} = 2a \frac{\left(\frac{1}{2^p}\right)^{\alpha}}{1-2^{-\alpha}} \leq 2a \frac{1}{1-2^{-\alpha}} (s_2 - s_1)^{\alpha}. \ \#$$

<u>Lemma 3.3.</u> $w(I_{\alpha,a,k,n}) \leq \sqrt{\frac{2}{\pi}} \frac{1}{a} 2^{n(\alpha - \frac{1}{2})} e^{-\frac{a^2}{2} \cdot 2^{n(1-2\alpha)}}$

Proof. Note that $I_{\alpha,a,k,n}$ is a cylinder set. Recall that $x(t)-x(s)$ is normally distributed with mean 0 and variance $t-s$, $(t > s)$. Therefore,

$$w(I_{\alpha,a,k,n}) = \frac{2}{\sqrt{2\pi \frac{1}{2^n}}} \int_{a(\frac{1}{2^n})^{\alpha}}^{\infty} e^{-\frac{\tau^2}{2\frac{1}{2^n}}} d\tau$$

$$= \sqrt{\frac{2}{\pi}} \sqrt{2^n} \int_{a(\frac{1}{2^n})^{\alpha}}^{\infty} e^{-\frac{\tau^2}{2\frac{1}{2^n}}} d\tau$$

$$= \sqrt{\frac{2}{\pi}} \int_{a(\frac{1}{2^n})^{\alpha - \frac{1}{2}}}^{\infty} e^{-\frac{\tau^2}{2}} d\tau.$$

But $$\int_b^{\infty} e^{-\frac{\tau^2}{2}} d\tau \leq \int_b^{\infty} \left(\frac{\tau}{b}\right) e^{-\frac{\tau^2}{2}} d\tau$$

$$= \frac{1}{b} e^{-\frac{b^2}{2}} \quad \text{for } b > 0.$$

Hence

$$w(I_{\alpha,a,k,n}) \leq \sqrt{\frac{2}{\pi}} \frac{1}{a} \cdot 2^{n(\alpha - \frac{1}{2})} e^{-\frac{a^2}{2} \cdot 2^{n(1-2\alpha)}}. \qquad \#$$

Lemma 3.4. For $\alpha > 0$ and $a > 0$, we have

$$w^*(H_\alpha[2a \frac{1}{1-2^{-\alpha}}]) \le \sqrt{\frac{2}{\pi}} \frac{1}{a} \sum_{k=0}^{\infty} 2^{k(\alpha+\frac{1}{2})} e^{-\frac{a^2}{2}\cdot 2^{k(1-2\alpha)}}.$$

Proof. Clearly, Lemma 3.2 says that

$$\bigcap_{n=0}^{\infty} \bigcap_{k=1}^{2^n} I^c_{\alpha,a,k,n} \subset H_\alpha[2a\frac{1}{1-2^{-\alpha}}]^c.$$

Hence $H_\alpha[2a \frac{1}{1-2^{-\alpha}}] \subset \bigcup_{n=0}^{\infty} \bigcup_{k=1}^{2^n} I_{\alpha,a,k,n}$.

Therefore,

$$w^*(H_\alpha[2a \frac{1}{1-2^{-\alpha}}]) \le \sum_{n=0}^{\infty} \sum_{k=1}^{2^n} w(I_{\alpha,a,k,n})$$

$$\le \sum_{n=0}^{\infty} \sum_{k=1}^{2^n} \sqrt{\frac{2}{\pi}} \frac{1}{a} \cdot 2^{n(\alpha-\frac{1}{2})} e^{-\frac{a^2}{2}\cdot 2^{n(1-2\alpha)}}$$

$$= \sqrt{\frac{2}{\pi}} \frac{1}{a} \sum_{n=0}^{\infty} 2^n \cdot 2^{n(\alpha-\frac{1}{2})} e^{-\frac{a^2}{2}\cdot 2^{n(1-2\alpha)}}$$

$$= \sqrt{\frac{2}{\pi}} \frac{1}{a} \sum_{n=0}^{\infty} 2^{n(\alpha+\frac{1}{2})} e^{-\frac{a^2}{2}\cdot 2^{n(1-2\alpha)}}$$

Remark. The above series is easily seen to be divergent for $\alpha \ge \frac{1}{2}$. It is convergent when $0 < \alpha < \frac{1}{2}$. This can be seen as follows. Let $\delta = \frac{1}{2} - \alpha$. Choose N so large that $N > \frac{1-\delta}{2\delta}$. Note that $e^{-x} \le \frac{N!}{x^N}$ for large x. Therefore the series is dominated by $\sum_{k=k_o}^{\infty} (2^{1-\delta-2\delta N})^k$ for some k_o.

Hence the series is convergent. However, we can show a little bit more.

Lemma 3.5. Let $a > o$ and $o < \alpha < \frac{1}{2}$. If I is a cylinder set contained in $H_\alpha [2a\frac{1}{1-2^{-\alpha}}]$ then $w(I) \leq \sqrt{\frac{2}{\pi}}\,\frac{1}{a}\,\frac{1}{1-2^{1-\delta}e^{-\frac{1}{2}a^2\delta}}$, where $\delta = \frac{1}{2} - \alpha$.

Remark. Note that $\lim_{a\to\infty} \frac{1}{a}\,\frac{1}{1-2^{1-\delta}e^{-\frac{1}{2}a^2\delta}} = 0$.

Proof. We use only a rough estimate, namely, $2^y \geq \frac{y}{2}$, $(y \geq 0)$.

Hence,
$$\sum_{k=0}^{\infty} 2^{k(1-\delta)} e^{-\frac{a^2}{2}\cdot 2^{2\delta k}} \leq \sum_{k=o}^{\infty} 2^{k(1-\delta)} e^{-\frac{a^2}{2}\delta k}$$
$$= \sum_{k=0}^{\infty} (2^{(1-\delta)}\cdot e^{-\frac{a^2\delta}{2}})^k$$
$$= \frac{1}{1-2^{1-\delta}e^{-\frac{1}{2}a^2\delta}}.$$
#

Theorem 3.1. (Wiener) w is σ-additive in the σ-field generated by R .

Proof. We need to show that if I_n is a decreasing sequence of cylinder sets with empty intersection then $\lim_{n\to\infty} w(I_n) = 0$.

Let
$$I_n = I_n(t_1^{(n)}, t_2^{(n)}, \ldots, t_{s_n}^{(n)}; E_n)$$
$$\equiv \{x \in C;\ (x(t_1^{(n)}), x(t_2^{(n)}), \ldots, x(t_{s_n}^{(n)})) \in E_n \subset R^{s_n}\}.$$

Step 1: Choose closed set $G_n \subset E_n$ such that $w(I_n \setminus K_n) < \frac{\varepsilon}{2^{n+1}}$, where $K_n = K_n(t_1^{(n)}, t_2^{(n)}, \ldots, t_{s_n}^{(n)}; G_n)$. Let

$$L_n = \bigcap_{j=1}^{n} K_j \in R. \quad \text{Then } L_n \subset K_n \subset I_n.$$

Hence $w(I_n) = w(I_n \setminus L_n) + w(L_n)$.

But $I_n \setminus L_n = I_n \setminus \bigcap_{j=1}^{n} K_j = \bigcup_{j=1}^{n} (I_n \setminus K_j) \subset \bigcup_{j=1}^{n} (I_j \setminus K_j)$,

so, $w(I_n \setminus L_n) \leq \sum_{j=1}^{n} \frac{\varepsilon}{2^{j+1}} \leq \frac{\varepsilon}{2}$.

Thus we have

$$w(I_n) \leq \varepsilon/2 + w(L_n) \text{ for all } n.$$

Step 2: We show that there exists n_o such that $w(L_n) < \varepsilon/2$ whenever $n \geq n_o$. (We then have $w(I_n) < \varepsilon$ whenever $n \geq n_o$. This means $\lim_{n\to\infty} w(I_n) = 0$, which is just what we wanted to prove.)

Let $b = 2a \frac{1}{1-2^{-\alpha}}$, where $0 < \alpha < \frac{1}{2}$. By Lemma 3.4 and Lemma 3.5 we can choose b large enough that $w(I) < \varepsilon/2$ whenever $I \subset H_\alpha[b]$.

Obviously, $H_\alpha[b]^c = \{x \in C;\ |x(t) - x(s)| \leq b|t-s|^\alpha\ \forall t,s\}$.

We will have proved our assertion in this step if we show that there exists n_0 s.t.

$$M_n \equiv L_n \cap H_\alpha[b]^c = \phi, \forall n \geq n_0.$$

Observe that $M_n \downarrow$ and $\bigcap_{n=1}^{\infty} M_n = \phi$. Suppose $M_n \neq \phi$ for all n. Pick up $x_n \in M_n$ for each n. Consider the sequence $\{x_n;\ n=1,2,\ldots\}$ in $C[0,1]$. $\{x_n\}$ is equi-continuous because $x_n \in H_\alpha[b]^c$. Moreover, $\{x_n(t);\ n=1,2,\ldots\} \subset \mathbb{R}$ is bounded for each t because

$|x_n(t)| \le bt^{\alpha}$. Therefore, by Ascoli-Arzela's theorem $\{x_n;\ n=1,2,\ldots\} \subset C[0,1]$ is pre-compact. Hence there exists a subsequence, still denoted by $\{x_n\}$ for the sake of convenience, such that $x_n \to x_o \in C[0,1]$ uniformly. Obviously, $x_o \in H_\alpha[b]^c$. Fix n_o, then $x_n \in M_{n_o}$ $\forall n \ge n_o$. Note that M_{n_o} is compact, so $x_o \in M_{n_o}$. Therefore, $x_o \in M_n$ for all n. Hence $x_o \in \bigcap_{n=1}^{\infty} M_n$. This is a contradiction because $\bigcap_{n=1}^{\infty} M_n = \phi$.

Finally, recall that M_n is decreasing and note that we have just seen that it is impossible to have $M_n \neq \phi$ for all n. Therefore $M_{n_o} = \phi$ for some n_o and obvously, $M_n = \phi$ for $n \ge n_o$. #

<u>Theorem 3.2</u>. (a) $w(C_\alpha) = 1$ if $0 < \alpha < \frac{1}{2}$

(b) $w(C_\alpha) = 0$ if $\alpha > \frac{1}{2}$

Proof. (a) $C_\alpha = B_\alpha$ by Lemma 3.1 (b)

$= H_\alpha^c$ by Lemma 3.1 (d)

$= \bigcup_{n=1}^{\infty} H_\alpha^c[a_n]$ by Lemma 3.1 (c),

where $a_n > 0$ and $a_n \to \infty$ as $n \to \infty$.

Hence, $w(C_\alpha) = \lim_{n\to\infty} w(H_\alpha^c[a_n]) = 1 - \lim_{n\to\infty} w(H_\alpha[a_n])$

$= 1 - 0$ by Lemma 3.4

$= 1.$

(b) Let $J_{\alpha,a,n} = \{x \in C;\ |x(\frac{k}{2^n}) - x(\frac{k-1}{2^n})| \le a(\frac{1}{2^n})^{\alpha}$ for all $k=1,2,\ldots,2^n\}$.

Clearly, $H_\alpha^c[a] \subset J_{\alpha,a,n}$ for all $n = 1,2,3,\ldots$.

Now, use the same idea as in the proof of Exercise 10 to see that the random variables $x(\frac{k}{2^n}) - x(\frac{k-1}{2^n})$, $k=1,2,\ldots 2^n$, are independent and each is normally distributed with mean 0 and variance $\frac{1}{2^n}$. Therefore,

$$w(J_{\alpha,a,n}) = \prod_{k=1}^{2^n} w\{x \in C;\ |x(\frac{k}{2^n}) - x(\frac{k-1}{2^n})| \leq a(\frac{1}{2^n})^\alpha\}$$

$$= \prod_{k=1}^{2^n} \int_{-a(\frac{1}{2^n})^\alpha}^{a(\frac{1}{2^n})^\alpha} \frac{1}{\sqrt{2\pi\cdot\frac{1}{2^n}}}\, e^{-\frac{\tau^2}{2\cdot\frac{1}{2^n}}}\, d\tau$$

$$= \prod_{k=1}^{2^n} \sqrt{\frac{2}{\pi}} \int_0^{a(\frac{1}{2^n})^{\alpha-\frac{1}{2}}} e^{-\frac{\tau^2}{2}}\, d\tau$$

$$\leq \prod_{k=1}^{2^n} \{\sqrt{\frac{2}{\pi}}\, a(\frac{1}{2^n})^{\alpha-\frac{1}{2}}\}$$

$$= (\sqrt{\frac{2}{\pi}}\, a(\frac{1}{2^n})^{\alpha-\frac{1}{2}})^{2^n}$$

$$= e^{2^n\{\log\sqrt{\frac{2}{\pi}}\, a - (\alpha-\frac{1}{2})n\log 2\}} \to 0 \quad \text{as } n \to \infty .$$

Hence $\lim_{n\to\infty} w(J_{\alpha,a,n}) = 0$ for each $a > 0$, $\alpha > \frac{1}{2}$.

Therefore, $w(H_\alpha^c[a]) = 0$ for each $a > 0$, and $\alpha > \frac{1}{2}$.

It follows from Lemma 3.1 that $w(C_\alpha) = 0$ for $\alpha > \frac{1}{2}$. #

We want to prove now that the σ-field generated by the cylinder sets is the Borel field of $C[0,1]$. By the following two exercises, we see that to show $\sigma[\mathcal{R}] = \mathcal{B}$ it is sufficient to prove that the closed unit ball is in $\sigma[\mathcal{R}]$.

<u>Exercise 12</u>. $C[0,1]$ with the sup norm is separable.

<u>Exercise 13</u>. Let X be a separable metric space. Then every open set in X is a countable union of open balls.

<u>Theorem 3.3</u>. $\{x; \|x\| \leq 1\} \in \sigma[\mathcal{R}]$.

Proof. It is easy to see that

$$\{x; \|x\| \leq 1\} = \bigcap_{n=1}^{\infty} \{x; |x(t)| \leq 1 \quad \forall t = \frac{k}{2^n} \quad k = 1,2,\ldots 2^n\}.$$

#

We now prove a special case of Kac's formula [25]. The proof here, being different from Kac's, is due to Ito [24].

Let $f \in L^2[0,1]$. Define $\phi_f : C[0,1] \to R$ by

$$\phi_f(x) = \int_0^1 x(t)\, f(t)dt.$$

<u>Lemma 3.6</u>. ϕ_f is normally distributed with mean 0 and variance $\int_0^1 \int_0^1 \min(t,s)\, f(t)\, f(s)\, dtds$.

Proof. Let $g(t)$, $0 \leq t \leq 1$, be a function of bounded variation, then the Stieltjes integral $\theta_g(x) \equiv \int_0^1 g(t)dx(t)$ exists for each x in $C[0,1]$. We will see later on (Theorem 5.1) that θ_g is normally distributed.

Let $f \in L^2[0,1]$. Define $g(t) = \int_t^1 f(s)ds$.

then

$$\phi_f(x) = \int_0^1 x(t)f(t)dt$$

$$= x(t)(-g(t))\Big|_0^1 + \int_0^1 g(t)dx(t)$$

$$= \int_0^1 g(t)dx(t)$$

$$= \theta_g(x).$$

Therefore, ϕ_f is normally distributed. Moreover,

$$E_w[\phi_f] = E_w \int_0^1 x(t)f(t)\,dt$$

$$= \int_0^1 E_w[x(t)]\,f(t)\,dt$$

$$= 0 \text{ (by Example 1),}$$

and

$$E_w[\phi_f^2] = E_w \int_0^1 \int_0^1 x(t)\,x(s)f(t)f(s)\,dt\,ds$$

$$= \int_0^1 \int_0^1 E_w[x(t)x(s)]f(t)f(s)\,dt\,ds$$

$$= \int_0^1 \int_0^1 \min(t,s)\,f(t)f(s)\,dt\,ds. \qquad \#$$

Theorem 3.4. (Kac) $E_w[e^{-\alpha\int_0^1 x(t)^2\,dt}] = \sqrt{\operatorname{sech}\sqrt{2\alpha}}$, $\alpha \geq 0$.

Proof. Step 1 : Note that $C[0,1] \subset L^2[0,1]$ (real-valued functions). Let $\{e_n(\cdot)\}$ be an orthonormal basis of $L^2[0,1]$, then

$$\int_0^1 x(t)^2\,dt = |x|^2 = \sum_{n=1}^{\infty} \langle x, e_n\rangle^2$$

$$= \sum_{n=1}^{\infty} \int_0^1 x(t)e_n(t)\,dt.$$

Hence

$$E_w[e^{-\alpha\int_0^1 x(t)^2dt}] = E_w[e^{-\alpha\sum_{n=1}^{\infty}<x,e_n>^2}]$$

$$= E_w[\prod_{n=1}^{\infty} e^{-\alpha<x,e_n>^2}]$$

$$= E_w[\prod_{n=1}^{\infty} e^{-\alpha(\int_0^1 x(t)e_n(t)dt)^2}]$$

$$= E_w[\prod_{n=1}^{\infty} e^{-\alpha\phi_{e_n}(x)^2}].$$

Step 2. Define an operator S of L^2 [0,1] by

$$S f(t) = \int_0^1 \min(t,s)f(s)\, ds.$$

It is easy to see that

$$E_w[\phi_f\ \phi_g] = <S f,g> , \quad f,g\in L^2\ [0,1].$$

Let $\{\lambda_n\}$ be eigenvalues of S and $\{e_n\}$ the corresponding normalized eigenvectors (which we take to be those in Step 1). Observe that ϕ_{e_n}'s are independent because:

$$E_w[\phi_{e_n}\ \phi_{e_m}] = <S e_n,e_m> = \lambda_n\ \delta_{n\,m}.$$

Hence,

$$E_w[e^{-\alpha\int_0^1 x(t)^2\, dt}] = \prod_{n=1}^{\infty} E_w[e^{-\alpha\phi_{e_n}(x)^2}]$$

$$= \prod_{n=1}^{\infty}(1 + 2\alpha <S e_n,e_n>)^{-\frac{1}{2}}$$

$$= \prod_{n=1}^{\infty}[(1 + 2\alpha\lambda_n)^{-1}]^{\frac{1}{2}}.$$

Step 3. We want to find eigenvalues of the operator S.

$$S\,f(t) = \int_0^1 \min(t,s)\, f(s)\, ds = \int_0^t s\, f(s)\, ds + t \int_t^1 f(s)\, ds.$$

Let $S\,f = \lambda f$, $\lambda \neq 0$.

Since $S\,f$ is continuous, hence $f = \frac{1}{\lambda} S\,f$ is continuous, and hence $S\,f$ is differentiable. Therefore,

$$\lambda f'(t) = t\,f(t) + \int_t^1 f(s)\, ds - t\,f(t) = \int_t^1 f(s)\, ds.$$

Since f is continuous, f' is differentiable. Hence

$$\lambda f'' = -f.$$

To find the initial conditions, note that $f(0) = S\,f(0) = 0$ and $f'(1) = \int_1^1 f(s)\, ds = 0$. Hence we have to solve the following equation

$$\lambda f'' + f = 0, \quad f(0) = f'(1) = 0.$$

It is easy to find the solutions, namely, $\lambda = \left(\frac{1}{(n-\frac{1}{2})\pi}\right)^2$ and $f(t) = \sqrt{2} \sin[(n - \tfrac{1}{2})\pi t]$.

Therefore,

$$E_w[e^{-\alpha\int_0^1 x(t)^2\, dt}] = \left\{\prod_{n=1}^{\infty} [1 + (\frac{\sqrt{2\alpha}}{(n-\frac{1}{2})\pi})^2]^{-1}\right\}^{1/2} = \sqrt{\operatorname{sech}\sqrt{2\alpha}}. \quad \#$$

Exercise 14. Prove that

$$E_w[e^{-\alpha\int_0^1 (x(t) - t\,x(1))^2\, dt}] = \sqrt{\sqrt{2\alpha}\,\operatorname{csch}\sqrt{2\alpha}}.$$

We give a brief discussion relating the solution of a partial differential equation to function space integral (i.e. Wiener integral in $C[0,1]$).

Definition 3.3. Donsker's delta function $\delta_{t,\xi}$, $(t > 0, \xi \in \mathbb{R})$, is defined formally by

$$\delta_{t,\xi}(x) = \frac{1}{2\pi} \int_{-\infty}^{\infty} e^{i\,y(x(t)-\xi)}\,dy, \quad x \in C[0,1]$$

<u>Lemma 3.7</u>. $E_w[\delta_{t,\xi}] = \frac{1}{\sqrt{2\pi t}}\, e^{-\frac{\xi^2}{2t}}$

Proof. $E_w[\delta_{t,\xi}] = \frac{1}{2\pi} \int_{-\infty}^{\infty} e^{-iy\xi} E_w(e^{iyx(t)})\,dy$

$$= \frac{1}{2\pi} \int_{-\infty}^{\infty} e^{-iy\xi} \cdot e^{-\frac{ty^2}{2}}\,dy$$

$$= \frac{1}{\sqrt{2\pi t}}\, e^{-\frac{\xi^2}{2t}}. \qquad \#$$

<u>Lemma 3.8</u>. $\int_{-\infty}^{\infty} f(\xi)\, E_w\{G(x)\,\delta_{t,\xi}(x)\}\,d\xi = E_w\{G(x) f(x(t))\}$.

Proof. $\text{L.H.S.} = E_w\{G(x) \int_{-\infty}^{\infty} f(\xi) \frac{1}{2\pi} \int_{-\infty}^{\infty} e^{iy(x(t)-\xi)}\,dy\,d\xi\}$

$= E_w\{G(x) f(x(t))\}$ (by the inversion formula of Fourier transforms). #

<u>Theorem 3.5</u>. (Donsker-Lions) The function

$$u(t,\xi) = E_w\{\delta_{t,\xi}(x)\, e^{-\int_0^t V(x(s))ds}\}$$

is a solution of the partial differential equation

$$u_t = \frac{1}{2} u_{\xi\xi} - V(\xi) u$$

$$u(t,\xi) \to 0 \text{ as } \xi \to \pm\infty$$

$$u(t,\xi) \to \delta(\xi) \text{ as } t \to 0, \text{ where } V \text{ is bounded below.}$$

Proof. Obviously,

$$e^{-\int_0^t V(x(s))\,ds} = 1 - \int_0^t V(x(\tau))\, e^{-\int_0^\tau V(x(s))ds}\,d\tau.$$

Hence

$$u(t,\xi) = E_W\{\delta_{t,\xi}(x)\} - \int_0^t E_W\{\delta_{t,\xi}(x)\, V(x(\tau))e^{-\int_0^\tau V(x(s))\,ds}\}d\tau$$

$$= \frac{1}{\sqrt{2\pi t}} e^{-\frac{\xi^2}{2t}} - \frac{1}{2\pi}\int_0^t d\tau \int_{-\infty}^{\infty} dy\, e^{-iy\xi} E_W\{V(x(\tau))e^{-\int_0^\tau V(x(s))ds}\, e^{iy\,x(t)}\}.$$

But,

$$E_W\{V(x(\tau))e^{-\int_0^\tau V(x(s))ds}\cdot e^{iy\,x(t)}\}$$

$$= E_W\{[V(x(\tau))e^{-\int_0^\tau V(x(s))ds}\cdot e^{iyx(\tau)}][e^{iy(x(t)-x(\tau))}]\}$$

$$= E_W\{V(x(\tau))\, e^{-\int_0^\tau V(x(s))ds}\cdot e^{iy\,x(\tau)}\}\; E_W\{e^{iy(x(t)-x(\tau))}\}$$

$$= e^{-\frac{t-\tau}{2}y^2} E_W\{V(x(\tau))e^{-\int_0^\tau V(x(s))ds}\, e^{iy\,x(\tau)}\}$$

$$= e^{-\frac{t-\tau}{2}y^2}\int_{-\infty}^{\infty} V(\eta)\, e^{iy\eta}\, E_W\{e^{-\int_0^\tau V(x(s))ds}\,\delta_{\tau,\eta}(x)\}d\eta$$

(by Lemma 3.8 with $f(\xi) = V(\xi)e^{iy\xi}$ and $G(x) = e^{-\int_0^\tau V(x(s))ds}$),

and $\int_{-\infty}^{\infty} e^{-iy\xi}\cdot e^{-\frac{t-\tau}{2}y^2}\cdot e^{iy\eta}\, dy = \frac{\sqrt{2\pi}}{\sqrt{t-\tau}}\, e^{-\frac{(\xi-\eta)^2}{2(t-\tau)}}$.

Therefore,

$$u(t,\xi) = \frac{1}{\sqrt{2\pi t}}\, e^{-\frac{\xi^2}{2t}} - \int_0^t\int_{-\infty}^{\infty} V(\eta)u(\tau,\eta)\,\frac{1}{\sqrt{2\pi(t-\tau)}}\, e^{-\frac{(\xi-\eta)^2}{2(t-\tau)}}\, d\eta\, d\tau$$

Now, it is very easy to check that $u(t,\xi)$ satisfies the partial differential equation in the theorem. #

To conclude this section, we mention Wiener measure w_c in $C[0,1]$ of variance $c > 0$.

Definition 3.4. Wiener measure of variance $c > 0$ is defined by

$$w_c(I)=\left[\prod_{k=1}^{n} 2\pi c(t_k-t_{k-1})\right]^{-1/2}\int_E e^{-\frac{1}{2c}\sum_{k=1}^{n}\frac{(u_k-u_{k-1})^2}{t_k-t_{k-1}}}du_1\ldots du_n,$$

where $t_o=0$, $u_o=0$ and

$$I=\{x\in C[0,1];(x(t_1),x(t_2),\ \ldots\ x(t_n))\in E\}.$$

Theorem 3.6. $\{w_c\}$ are Borel measures in $C[0,1]$. Moreover, $\{w_c;c > 0\}$ form a semi-group acting in the Banach space of bounded, uniformly continuous functions defined in $C[0,1]$.

Proof. The first assertion is obvious since the argument in showing the σ-additivity of $w=w_1$ applies equally to w_c. The second assertion is left as Exercise 15. #

§4. Abstract Wiener Space.

Let H be a real separable Hilbert space with norm $|\cdot| = \sqrt{\langle\cdot,\cdot\rangle}$. $\mathcal{F}$ will denote the partially ordered set of finite dimensional orthogonal projections P of H. ($P > Q$ means $P(H) \supset Q(H)$ for $P,Q \in \mathcal{F}$).

A subset E of H of the following form is called a cylinder set,

$$E = \{x \in H;\ Px \in F\},$$

where $P \in \mathcal{F}$ and F is a Borel subset of PH. $\mathcal{R}$ will denote the collection of cylinder sets. It is easy to see that $\mathcal{R}$ is a field, but is not a σ-field.

Definition 4.1. A function f in H of the form $f(x) = \phi(Px)$ is called a cylinder function, where ϕ is a Borel function defined in PH and $P \in \mathcal{F}$.

It is easy to see that f is a cylinder function iff it is of the form $f(x) = \psi(\langle x,e_1\rangle,\cdots,\langle x,e_n\rangle)$, where $\{e_n\}$ is an orthonormal set and ψ is a Borel function defined in $\mathbb{R}^n$.

Definition 4.2. The Gauss measure in H is the set function μ from $\mathcal{R}$ into $[0,\infty)$ defined as follows: If $E = \{x \in H;\ Px \in F\}$ then

$$\mu(E) = \left(\frac{1}{\sqrt{2\pi}}\right)^n \int_F e^{-\frac{|x|^2}{2}}\,dx,$$

where $n = \dim PH$ and dx is the Lebesgue measure of PH. Obviously, μ is finitely additive, but we have the following.

Proposition 4.1. μ is not σ-additive.

Proof. Let $\{e_n\}$ be an orthonormal basis of H. Let

$$E_n = \{x \in H;\ |<x,e_k>| \le n,\quad k = 1,2,\cdots,a_n\}.$$

Clearly, $H = \bigcup_{n=1}^{\infty} E_n$ no matter how we choose $a_n \nearrow \infty$.

Now,

$$\mu(E_n) = \left(\frac{1}{\sqrt{2\pi}}\right)^{a_n} \underbrace{\int_{-n}^{n} \cdots\cdots \int_{-n}^{n}}_{a_n} e^{-\frac{1}{2}\sum_{j=1}^{a_n} x_j^2}\, dx_1 \ldots dx_{a_n}$$

$$= \left(\frac{1}{\sqrt{2\pi}} \int_{-n}^{n} e^{-\frac{u^2}{2}}\, du\right)^{a_n}.$$

For each n choose a_n so large that $\mu(E_n) < \frac{1}{2^{n+1}}$. Obviously, we may choose a_n in such a way that a_n increases to ∞. Then $\sum_{n=1}^{\infty} \mu(E_n) < 1/2$. However,

$$H = \{x \in H;\ <x,e_1> \in \mathbb{R}\} \Rightarrow \mu(H) = \frac{1}{\sqrt{2\pi}} \int_{-\infty}^{\infty} e^{-\frac{x^2}{2}}\, dx = 1.$$

Thus μ is not σ-additive. #

Take an orthonormal basis $\{e_n\}_{n=1}^{\infty}$ of H. Define a Borel measure $\mu_{e_1,\ldots,e_n}$ in $\mathbb{R}^n$ by

$$\mu_{e_1,\ldots,e_n}(F) = \mu\{x \in H;\ (<x,e_1>,\ldots,<x,e_n>) \in F\}.$$

Obviously, $\{\mu_{e_1,\ldots,e_n};\ n = 1,2,3,\ldots\}$ is a consistent family of probability measures. Therefore, by Kolmogorov's theorem, there exist a probability space (Ω,m) and random variables $\xi_1, \xi_2, \ldots.$ such that for any n,

$$m\{\omega; (\xi_1(\omega),\xi_2(\omega),\cdots,\xi_n(\omega)) \in F\}$$

$$= \mu_{e_1,e_2,\ldots,e_n}(F) = \mu\{x \in H; (\langle x,e_1\rangle, \langle x,e_2\rangle,\ldots,\langle x,e_n\rangle) \in F\}.$$

Proposition 4.2. $\{\xi_n\}$ is an independent sequence of Gaussian variables. Each is normally distributed with mean 0 and variance 1.

Proof. $m\{\omega; \xi_j(\omega) < a\} = \mu\{x \in H; \langle x,e_j\rangle < a\}$

$$= \frac{1}{\sqrt{2\pi}} \int_{-\infty}^{a} e^{-\frac{u^2}{2}} du, \; j = 1,2,\ldots .$$

Hence ξ_j's are all normally distributed with mean 0 and variance 1. Moreover, suppose $j \neq k$,

$$E_m[\xi_j \; \xi_k] = \int\int_{\mathbb{R}^2} uv \; \mu_{e_j,e_k}(dudv)$$

$$= \int\int_{\mathbb{R}^2} uv. \frac{1}{2\pi} e^{-\frac{u^2+v^2}{2}} dudv$$

$$= (\int_{\mathbb{R}} \frac{1}{\sqrt{2\pi}} u\, e^{-\frac{u^2}{2}} du)^2 = 0.$$

Hence ξ_j and ξ_k are independent for $j \neq k$. #

Notation . $\xi_j = n(e_j)$, $j = 1,2,3,\ldots .$ n stands for the normal distribution introduced by Segal [41].

Let $h \in H$. Using Proposition 4.2, we see that the series of random variables $\sum_{j=1}^{\infty} \langle h,e_j\rangle\, n(e_j)$ converges in

$L^2(\Omega,m)$ to a unique random variable, which we denote by $n(h)$.

<u>Corollary 4.1.</u> (a) $n(h)$ is normally distributed with mean 0 and variance $|h|^2$.

(b) $E_m[n(h)n(k)] = \langle h,k\rangle$, $h,k \in H$. Hence if $h \perp k$ then $n(h)$ and $n(k)$ are independent.

Proof. (a)
$$E_m[e^{in(h)}] = \lim_{N\to\infty} E_m[e^{i\sum_{j=1}^{N}\langle h,e_j\rangle\, n(e_j)}]$$
$$= \lim_{N\to\infty} \prod_{j=1}^{N} E_m[e^{i\langle h,e_j\rangle n(e_j)}]$$
$$= \lim_{N\to\infty} \prod_{j=1}^{N} e^{-\frac{1}{2}\langle h,e_j\rangle^2}$$
$$= \lim_{N\to\infty} e^{-\frac{1}{2}\sum_{j=1}^{N}\langle h,e_j\rangle^2}$$
$$= e^{-\frac{1}{2}|h|^2}.$$

(b)
$$|h+k|^2 = E_m[n(h+k)^2] \quad \text{(by (a))}$$
$$= E_m[(n(h)+n(k))^2] \quad \text{(by Exercise 16 below)}$$
$$= |h|^2 + 2\,E_m[n(h)n(k)] + |k|^2,$$

whence (b) follows immediately. #

<u>Exercise 16.</u> Show that $n : H \to L^2(\Omega,m)$ is linear.

<u>Definition 4.3.</u> Let f be a cylinder function given by

$$f(x) = \phi(\langle x,e_{j_1}\rangle, \cdots, \langle x,e_{j_s}\rangle).$$

Define a random variable associated with f by

$$\tilde{f}(\omega) = \phi(n(e_{j_1})(\omega),\cdots,n(e_{j_s})(\omega)), \ \omega \in \Omega.$$

Remark . $\tilde{f}$, in general, is not in $L^2(\Omega,m)$. Therefore, we will use a weaker convergence, i.e., convergence in probability. Recall that a sequence of random variables g_n converges in probability to g if given any $\varepsilon > 0$, there exists n_0 such that

$$m\{\omega;\ |g_n(\omega) - g(\omega)| > \varepsilon\} < \varepsilon \quad \text{whenever } n \geq n_0.$$

Question : Let $\{e_k\}$ be an orthonormal basis of H and P_n the projection onto the span of $\{e_1,\ldots e_n\}$. Let $f_n(x) = |P_n x|^2$. Obviously, f_n is a cylinder function and $\lim_{n\to\infty} f_n(x) = |x|^2$ for each x in H. Does $\tilde{f}_n$ converge in probability ?

Answer: No. To see this, observe that when $n \geq k$

$$m\{\omega;\ |\tilde{f}_n(\omega) - \tilde{f}_k(\omega)| > \varepsilon\}$$

$$= \mu\{x \in H;\ ||P_n x|^2 - |P_k x|^2| > \varepsilon\}$$

$$= \mu\{x \in H;\ |(P_n - P_k)x|^2 > \varepsilon\}$$

$$= 1 - \mu\{x \in H;\ |(P_n - P_k)x|^2 \leq \varepsilon\}.$$

But $\{y_1^2+\cdots\cdots+y_j^2 \leq \varepsilon\} \subset \{|y_1| \leq \sqrt{\varepsilon},\ldots\ldots,|y_j| \leq \sqrt{\varepsilon}\}$.

Hence

$$m\{\omega;\ |\tilde{f}_n(\omega) - \tilde{f}_k(\omega)| > \varepsilon\}$$

$$\geq 1 - \left(\frac{1}{\sqrt{2\pi}} \int_{-\sqrt{\varepsilon}}^{\sqrt{\varepsilon}} e^{-\frac{u^2}{2}} du\right)^{n-k} \to 1 \text{ when } n-k \to \infty.$$

<u>Definition 4.4</u>. A semi-norm $||\cdot||$ in H is called <u>measurable</u> if for every $\varepsilon > 0$, there exists $P_o \in \mathcal{F}$ such that

$$\mu\{||\, P\, x\, || > \varepsilon\} < \varepsilon \qquad \forall\ P \perp P_o \text{ and } P \in \mathcal{F}.$$

Remark. The above argument shows that if $\dim(H) = \infty$ then $|\cdot|$ is not a measurable norm.

<u>Exercise 17</u>. Let $D \in \mathcal{L}_{(2)}(H)$. Define $||\, x\, || = |D\, x|$, $x \in H$. Show that $||\cdot||$ is a measurable semi-norm.

<u>Lemma 4.1</u>. If $||\cdot||$ is a measurable semi-norm, then the net $||\, P\, x\, ||^{\sim}$ converges in probability on Ω as $P \to I$ strongly through $\mathcal{F}$. (The limit will be denoted by $||\, x\, ||^{\sim}$).

Remark. It is easy to see that if we want the conclusion in the lemma then we must have the condition in Definition 4.4.

Proof. <u>Step 1</u>. Choose an increasing sequence $\{P_n\} \subset \mathcal{F}$ such that

$$\mu\{x;\ ||P\, x|| > \frac{1}{n}\} < \frac{1}{n} \quad \text{whenever } P \in \mathcal{F} \text{ and } P \perp P_n.$$

Let $\varepsilon > 0$ be given, choose N so large that $\frac{1}{N} < \varepsilon$. Then, when $n, m \geq N$, $P_n - P_m \perp P_N$ and so

$$\mu\{x;\ ||(P_n - P_m)\, x\, || > \frac{1}{N}\} < \frac{1}{N}.$$

Obviously,

$$\mu\{x;\ |\ \|P_n x\| - \|P_m x\|\ | > \varepsilon\} \leq \mu\{x; \|(P_n - P_m) x\| > \varepsilon\}$$

$$\leq \mu\{x;\ \|(P_n - P_m) x\| > \tfrac{1}{N}\}$$

$$< \tfrac{1}{N} < \varepsilon .$$

Hence $\| P_n x \|^{\sim}$ converges in probability to, g say, as

$n \to \infty$.

<u>Step 2</u>. Suppose $P > P_N$ and $P \in \mathcal{F}$, we have

$$m\{\omega; |g(\omega) - \|Px\|^{\sim}(\omega)| > 2\varepsilon\} \leq m\{\omega; |g(\omega) - \|P_N x\|^{\sim}(\omega)| > \varepsilon\}$$

$$+ m\{\omega; |\ \|P_N x\|^{\sim}(\omega) - \|P x\|^{\sim}(\omega)| > \varepsilon\}.$$

But, $m\{\omega;\ |g(\omega) - \|P_N x\|^{\sim}(\omega)| > \varepsilon\} < \varepsilon$,

and $m\{\omega;\ |\ \|P_N x\|^{\sim}(\omega) - \|P x\|^{\sim}(\omega)| > \varepsilon\}$

$$= \mu\{x \in H;\ |\ \|P_N x\| - \|P x\|\ | > \varepsilon\}$$

$$\leq \mu\{x \in H;\ \|(P_N - P) x\| > \varepsilon\}$$

$$\leq \mu\{x \in H;\ \|(P - P_N) x\| > \tfrac{1}{N}\}$$

$$\leq \tfrac{1}{N} < \varepsilon .$$

Hence, $m\{\omega;\ |g(\omega) - \|P x\|^{\sim}(\omega)| > 2\varepsilon\} < 2\varepsilon$.

This means that the net $||P x||^{\sim}$ converges in probability to g as $P \to I$ strongly through the directed set F. #

Lemma 4.2. Let $||\cdot||$ be a measurable semi-norm. Then there exists a constant c such that $||x|| \leq c|x|$ for all x in H.

Proof. Let a be a number such that

$$2\frac{1}{\sqrt{2\pi}}\int_a^\infty e^{-\frac{u^2}{2}}du = \frac{1}{2}.$$

Choose $P_o \in F$ such that

$$\mu\{||Px|| > \tfrac{1}{2}\} < \tfrac{1}{2}, \text{ whenever } P \in F \text{ and } P \perp P_o.$$

Since P_oH is finite dimensional, there exists a constant α such that

$$||y|| \leq \alpha |y| \qquad \text{for all } y \in P_oH.$$

On the other hand, if $z \neq 0$ and $z \in (P_oH)^{\perp}$, we define

$$P_z x = \langle x, \frac{z}{|z|}\rangle \frac{z}{|z|},$$

then $P_z \in F$ and $P_z \perp P_o$. Therefore,

$$\mu\{||P_z x|| > \tfrac{1}{2}\} < \tfrac{1}{2},$$

i.e. $$\mu\{|\langle x, \frac{z}{|z|}\rangle| > \frac{|z|}{2||z||}\} < \frac{1}{2}$$

i.e. $$2\frac{1}{\sqrt{2\pi}}\int_{\frac{|z|}{2||z||}}^{\infty} e^{-\frac{u^2}{2}}du < \frac{1}{2}.$$

Hence, $\frac{|z|}{2||z||} > a$, or $|| z || < \frac{1}{2a} | z |$.

Finally, for any $x \in H$, x can be written uniquely as $x = y + z$, where $y \in P_o H$ and $z \in (P_o H)^{\perp}$.

Thus,

$$\begin{aligned} || x ||^2 \leq (||y|| + ||z||)^2 &\leq 2(|| y ||^2 + || z ||^2) \\ &\leq 2(\alpha^2 | y |^2 + \frac{1}{4a^2} | z |^2) \\ &\leq 2(\alpha^2 + \frac{1}{4a^2})(|y|^2 + |z|^2) \\ &= 2(\alpha^2 + \frac{1}{4a^2}) | x |^2. \end{aligned}$$

Hence

$$|| x || \leq c | x |, \text{ where } c = \sqrt{2} (\alpha^2 + \frac{1}{4a^2})^{1/2}. \quad \#$$

Lemma 4.3. Let $||\cdot||$ be a measurable semi-norm on H and A a bounded linear operator of H. Then $||| x ||| = ||A x||$ is also a measurable semi-norm. Moreover, for any $\varepsilon > 0$,

$$m\{\omega \in \Omega; |||x|||^{\sim} (\omega) > || A || \varepsilon\} \leq m\{\omega \in \Omega ; || x ||^{\sim}(\omega) > \varepsilon\}$$

Proof. See [16, page 383].

In view of Lemma 4.2, any measurable semi-norm is weaker than the norm of H. Let us take a measurable norm $||\cdot||$ and complete H with respect to $||\cdot||$. Observe that H is not complete with respect to $||\cdot||$ unless H is finite dimensional. If it were complete then $||\cdot||$ would be equivalent to $|\cdot|$ by the open mapping theorem. And $|\cdot|$ would be measurable. However, by the remark following Definition 4.4, $|\cdot|$ is not measurable.

Notation. B = the completion of H with respect to $||\cdot||$. i will denote the inclusion map of H into B. The triple (i,H,B) is called an abstract Wiener space. Later on, we will see that C[0,1] arises in this way. C [0,1] will be referred to as the classical Wiener space.

Exercise 18. Let H and H_o be as given in the end of §2. Let i be the inclusion map of H_o into H. Show that (i,H_o, H) is an abstract Wiener space.

Remark. Later on, we will see that in case B is a Hilbert space then (i,H,B) arises in the way given in the above exercise.

Now, recall the Gauss measure μ in H. Proposition 4.1 shows that μ is not σ-additive in R. Regarding $y \in B^*$ as an element of $H^* \equiv H$ by restriction, we can embed B^* in H. Define

$$\tilde{\mu}\{x \in B; ((x,y_1),\dots,(x,y_n)) \in E\} = \mu\{x \in H; (\langle x,y_1\rangle,\dots\langle x,y_n\rangle) \in E\},$$

where y_j's are in B^* and (,) denotes the natural pairing between B and B^*.

Notation. A set of the form $\{ x \in B; ((x,y_1),\dots(x,y_n) \in E \}$ is called a cylinder set in B. Let R_B denote the collection of cylinder sets in B.

Theorem 4.1. (Gross) $\tilde{\mu}$ is σ-additive in the σ-field generated by R_B.

Remark. Later on, we will show that the σ-field generated by R_B is the Borel field of B. The proof of this theorem depends on the following lemmas.

Lemma 4.4. Let $||\cdot||$ be a measurable semi-norm on H. Let $\{a_n; n = 0,1,2,...\}$ be a sequence of positive numbers. Then there exists a sequence $\{Q_n; n=0,1,2,...\} \subset F$ such that

(a) $Q_j Q_k = \delta_{jk} Q_j$, $\sum_{n=0}^{\infty} Q_n = I$ strongly,

and (b) $|| x ||_0 = \sum_{n=0}^{\infty} a_n || Q_n x ||$ converges for each $x \in H$ and $||\cdot||_0$ is a measurable semi-norm.

Remark. Observe that $||\cdot||_0$ is a norm when $||\cdot||$ is a norm.

Proof. From the definition of measurability of $||\cdot||$, we can choose, for each $n \geq 1$, $P_n \in F$ such that $P_n \nearrow I$ strongly and

$\mu \{ || Px || > \frac{1}{a_n 2^n} \} < \frac{1}{2^n}$, whenever $P \in F$ and $P \perp P_n$.

Define $Q_0 = P_1$ and $Q_n = P_{n+1} - P_n$, $n \geq 1$.

Obviously we have (a).

Let $\alpha_n > 0$ be chosen such that

$$2 \frac{1}{\sqrt{2\pi}} \int_{\alpha_n}^{\infty} e^{-\frac{u^2}{2}} du = \frac{1}{2^n}, \quad n = 1,2,3,\ldots .$$

Let $z \in H$. Then since $Q_n z \perp P_n$, we have

$$\mu\{x \in H;\ \|P_{Q_n z}\, x\| > \frac{1}{a_n 2^n}\} < \frac{1}{2^n},$$

where $P_{Q_n z}$ is the orthogonal projection onto the span of $Q_n z$. By the same argument as in the proof of Lemma 4.2, we have

$$a_n \|Q_n z\| \leq \frac{1}{2^n \alpha_n} |Q_n z| \qquad n=1,2,3,\ldots .$$

On the other hand, it is obvious that there exists α_0 such that $a_0 \|Q_0 z\| \leq \frac{1}{\alpha_0} |Q_0 z|$ for all $z \in H$.

Hence

$$\sum_{n=0}^{\infty} a_n \|Q_n z\| \leq \sum_{n=0}^{\infty} \frac{1}{2^n \alpha_n} |Q_n z|$$

$$\leq \left(\sum_{n=0}^{\infty} \left(\frac{1}{2^n \alpha_n}\right)^2\right)^{1/2} \left(\sum_{n=0}^{\infty} |Q_n z|^2\right)^{1/2}$$

$$= \left(\sum_{n=0}^{\infty} \left(\frac{1}{2^n \alpha_n}\right)^2\right)^{1/2} |z|.$$

Observe that $\sum_{n=0}^{\infty}\left(\frac{1}{2^n \alpha_n}\right)^2$ is convergent because $\alpha_n \geq 1$ for all large n. Hence $\sum_{n=0}^{\infty} a_n \|Q_n z\|$ converges for each $z \in H$ to, $\|z\|_0$ say. Evidently, $\|\cdot\|_0$ is a semi-norm in H.

We have to show that $\|\cdot\|_0$ is measurable. Let $\varepsilon > 0$ be given. Choose N so large that $\frac{1}{2^{N-1}} < \varepsilon$. Suppose $P \in \mathcal{F}$ and $P \perp P_N$. Then

$$m\{\|Px\|_0 > \varepsilon\} = m\{\sum_{n=0}^{\infty} a_n \|Q_n P x\| > \varepsilon\}$$

$$= m\{\sum_{n=N}^{\infty} a_n \|Q_n P x\| > \varepsilon\}$$

$$\leq m\{\sum_{n=N}^{\infty} a_n \|Q_n P x\| > \frac{1}{2^{N-1}}\}$$

$$\leq \sum_{n=N}^{\infty} m\{a_n \|Q_n P x\| > 1/2^n\}$$

$$= \sum_{n=N}^{\infty} m\{\|Q_n P x\| > 1/a_n 2^n\}.$$

Now, apply Lemma 4.3. to the measurable semi-norm $\|Q_n x\|$ and the operator P to conclude that

$$m\{\|Q_n P x\| > 1/a_n 2^n\} \leq m\{\|Q_n x\| > 1/a_n 2^n\}.$$

Therefore, $m\{\|Px\|_0 > \varepsilon\} \leq \sum_{n=N}^{\infty} m\{\|Q_n x\| > 1/a_n 2^n\}$

$$\leq \sum_{n=N}^{\infty} 1/2^n = 1/2^{N-1} < \varepsilon.$$

Hence $\|\cdot\|_0$ is measurable. #

<u>Corollary 4.2.</u> Let (i,H,B) be an abstract Wiener space. Then there exist another abstract Wiener space (i_0, H, B_0) and an increasing sequence $\{P_n\} \subset F$ converging strongly to the identity in H such that (a) B_0-norm is stronger than B-norm (hence $B_0 \subset B$), (b) each P_n extends by continuity to a projection $\tilde{P}_n$ of B_0, and (c) $\tilde{P}_n$ converges strongly to the identity in B_0 (w.r.t. B_0-norm).

Remark. Note that the conclusion implies the existence of a Schauder basis for B_o. In view of Theorem 4.4. below and the nonexistence of a Schauder basis of a real separable Banach space proved by Enflo, we can not hope to have $B_o = B$ in general.

Proof. Let $||\cdot||$ denote B-norm. Applying Lemma 4.4 to a sequence $a_n \geq 1$ for all $n \geq 0$, we see that there exists a sequence $\{Q_n;\ n=0,1,\ldots\} \subset F$ such that $Q_j Q_k = \delta_{jk} Q_j$, $\Sigma Q_n = I$ strongly and $|| x ||_o = \sum_{n=0}^{\infty} a_n || Q_n x ||$ is a measurable norm in H. Let B_o be the completion of H with respect to $||\cdot||_o$ and i_o the inclusion map from H into B_o. Then (i_o, H, B_o) is an abstract Wiener space. Let $x \in H$, we have

$$|| x || = \lim_{n\to\infty} ||Q_0 x + \cdots\cdots + Q_n x ||$$

$$\leq \lim_{n\to\infty}(|| Q_0 x|| + \cdots + || Q_n x ||)$$

$$\leq \lim_{n\to\infty}(a_0|| Q_0 x ||+ \cdots + a_n|| Q_n x||).$$

Hence, $|| x || \leq || x ||_o$ for all x in H and we have (a).

Define

$$P_n = Q_0 + \cdots\cdots + Q_{n-1},\ n \geq 1\ .$$

Clearly, $\{P_n\}$ is an increasing sequence converging strongly to the identity in H. Moreover, if $x \in H$, then

$$\| P_n x \|_o = \sum_{j=0}^{\infty} a_j \| Q_j P_n x \|$$

$$= \sum_{j=0}^{n} a_j \|Q_j x \|$$

$$\leq \| x \|_o .$$

Thus for each n, we have $\|P_n x \|_o \leq \| x \|_o$ for all x in H. Hence by continuity, each P_n extends to a bounded operator $\tilde{P}_n$ of B_o. Moreover, it is easy to see that each $\tilde{P}_n$ is a projection of B_o and $\|\tilde{P}_n \|_{B_o, B_o} \leq 1$. Thus we have (b). To see (c), first observe that if $x \in H$ then

$$\| P_n x - x \|_o = \sum_{j=0}^{\infty} a_j \|Q_j (P_n x - x)\|$$

$$= \sum_{j=n}^{\infty} a_j \|Q_j x \|$$

$$\to 0 \text{ as } n \to \infty$$

because $\sum_{j=0}^{\infty} a_j \|Q_j x \|$ is convergent. Now if $y \in B_o$, we choose a sequence $\{x_k\}$ in H such that $\|x_k - y\|_o \to 0$ as $k \to \infty$. Then

$$\|\tilde{P}_n y - y\|_o$$

$$\leq \|\tilde{P}_n y - \tilde{P}_n x_k\|_o + \|\tilde{P}_n x_k - x_k\|_o + \|x_k - y\|_o$$

$$\leq \|y - x_k\|_o + \|P_n x_k - x_k\|_o + \| x_k - y \|_o$$

It follows immediately that $\|P_n y - y \|_o \to 0$ as $n \to \infty$. #

Lemma 4.5. Let $||\cdot||$ be a measurable norm on H and B the completion of H with respect to $||\cdot||$. Then there exists a measurable norm $||\cdot||_o$ on H such that for each $r > 0$, $\{x \in H; \ ||x||_o \leq r\}$ is precompact in B.

Proof. Let $\{a_n\}$ be a sequence of positive numbers such that $\sum_{n=1}^{\infty} \frac{1}{a_n} < \infty$. By Lemma 4.4. there exists a sequence $\{Q_n\} \subset F$ such that Q_n's are mutually orthogonal, $\sum_{n=1}^{\infty} Q_n = I$ strongly and

$$||x||_o = \sum_{n=1}^{\infty} a_n ||Q_n x||$$

is a measurable norm on H. We show that $||\cdot||_o$ is a desired one. Let $r > 0$. To show that $\{x \in H; \ ||x||_o \leq r\}$ is precompact in B, it is sufficient to show if $||x_n||_o \leq r$, $n = 1,2,\ldots.$ then $\{x_n\}$ has a subsequence which is Cauchy with respect to $||\cdot||$.

For each $k = 1,2,3,\ldots.$, consider the sequence $\{Q_k x_n\}_{n=1}^{\infty}$. Since $a_k ||Q_k x_n|| \leq ||x_n||_o \leq r$, we have

$$||Q_k x_n|| \leq \frac{r}{a_k}, \quad \text{for all } n.$$

It follows that $\{Q_k x_n\}$ has a convergent subsequence. By diagonalization, we conclude that there exists a subsequence of x_n, still denoted by x_n, such that $\{Q_k x_n\}$ is Cauchy with respect to $||\cdot||$ for all k.

Now, by Lemma 4.2, if $y_n \to y$ in $|\cdot|$ then $y_n \to y$ in $||\cdot||$. Therefore, for all y in H,

$$||y|| = \lim_{n\to\infty} ||Q_1 y + \cdots + Q_n y|| .$$

Thus

$$||y|| \leq \sum_{n=1}^{\infty} ||Q_n y||, \quad y \in H.$$

In particular, we have

$$||x_n - x_m|| \leq \sum_{k=1}^{\infty} ||Q_k x_n - Q_k x_m|| .$$

Observe that the above series is dominated by $\sum_{k=1}^{\infty} 2r \frac{1}{a_k} < \infty$. Moreover, each term goes to 0 when n, m $\to \infty$. Therefore, $\lim_{n,m\to\infty} ||x_n - x_m|| = 0$. That is, $\{x_n\}$ is Cauchy w.r.t. $||\cdot||$. #

Lemma 4.6. Let $||\cdot||$ be a measurable semi-norm on H. Then there exists $K \in \mathcal{K}(H)$ and a measurable semi-norm $||\cdot||_o$ such that $||x|| \leq ||Kx||_o$ for all $x \in H$.

Proof. Let $\{a_n\}$ be a sequence such that $a_n \to \infty$ as $n \to \infty$. Choose Q_n's as in Lemma 4.4 such that

$$||x||_o = \sum_{n=1}^{\infty} a_n ||Q_n x||$$

is a measurable semi-norm in H.

Define an operator $K : H \to H$ by

$$Kx = x/a_n, \text{ when } x \in Q_n H.$$

Clearly, $K \in \mathcal{K}(H)$ (See Exercise 3). Moreover, for $x \in H$,

$$\|x\| \le \sum_{n=1}^{\infty} \|Q_n x\|$$

$$\le \sum_{n=1}^{\infty} a_n \|Q_n K x\|$$

$$= \|K x\|_o$$ #

Corollary 4.3. Let $\|\cdot\|$ be a measurable semi-norm. Then there is a compact operator K of H such that $\|x\| \le |Kx|$ for x in H.

Proof. By Lemma 4.2. and Lemma 4.6. #

Let $y \in B^*$. By Lemma 4.2, the restriction of y to H is a continuous linear functional of H, i.e., $y \in H^*$. Therefore $B^* \subset H^*$. By the Riesz representation theorem we have $H^* \approx H$. Therefore we have $B^* \subset H \subset B$. We will use (,) to denote the natural pairing between B^* and B and < , > the inner product of H. By the above identifications we have

$$(x,y) = \langle x,y \rangle \quad \text{whenever } x \in B^* \text{ and } y \in H.$$

Example 1. When $H = \ell_2$ and $\|x\| = \left(\sum_{n=1}^{\infty} \frac{1}{n^2} x_n^2\right)^{1/2}$ for $x = (x_1, x_2 \cdots\cdot)$.

Then $B = \{(x_1,x_2,\ldots);\ \Sigma \frac{1}{n^2} x_n^{\ 2} < \infty\}$ and

$$B^* = \{(x_1,x_2,\ldots);\ \Sigma n^2 x_n^{\ 2} < \infty\}.$$

Exercise 19. Show that $B^* = \{ x \in H;\ \sup_{\substack{y\in H \\ y\neq 0}} \frac{|<x,y>|}{||y||} < \infty \}$. (This gives a criterion to see when an element of H is actually in B*.)

Proof of Theorem 4.1.

Step 1. Obviously, to show that $\tilde{\mu}$ is σ-additive in the σ-field generated by R_B, it is sufficient to show that whenever $B = \bigcup_{n=1}^{\infty} T_n$, where T_n's are $||\cdot||$-open cylinder sets, then we have $\sum_{n=1}^{\infty} \tilde{\mu}(T_n) \geq 1$.

In Step 2 we will show that for any $\varepsilon > 0$ we have a compact set C_ε such that $\tilde{\mu}(T) < \varepsilon$ for all $T \in R_B$ and $T \cap C_\varepsilon = \phi$. Assuming this, we prove σ-additivity of $\tilde{\mu}$ as follows : Let $\varepsilon > 0$ be given, Choose C_ε as above. Then if $B = \bigcup_{n=1}^{\infty} T_n$ as above, we have $C_\varepsilon \subset T_1 \cup T_2 \cup \cdots \cup T_N$ for some N. Hence

$$\sum_{n=1}^{\infty} \tilde{\mu}(T_n) \geq \sum_{n=1}^{N} \tilde{\mu}(T_n)$$

$$\geq \tilde{\mu}\Big(\bigcup_{n=1}^{N} T_n\Big)$$

$$\geq 1 - \tilde{\mu}\Big(B \setminus \bigcup_{n=1}^{N} T_n\Big)$$

$$\geq 1 - \varepsilon.$$

It follows immediately that $\sum_{n=1}^{\infty} \tilde{\mu}(T_n) \geq 1$ since $\varepsilon > 0$ is arbitrary.

<u>Step 2.</u> To show what we assume in Step 1.

Pick up one measurable norm $||\cdot||_o$ on H as in Lemma 4.5. Choose $r > o$ such that

$m\{\omega \in \Omega;\ ||\ x\ ||_o^{\sim}(\omega) > r\} < \varepsilon$, (see Lemma 4.1. for the notation $||\cdot||_o^{\sim}$).

Let C_ε = the closure of $\{x \in H;\ ||\ x\ ||_o \leq r\}$ in B.

By Lemma 4.5. C_ε is compact in B.

Now, suppose $T \in R_B$ and $T \cap C_\varepsilon = \phi$. T can be written in the form $T = \{x \in B;\ ((x,y_1),\cdots\cdots,(x,y_n) \in E\}$, where E is a Borel subset of $\mathbb{R}^n$ and $y_j \in B^*$ $j=1,2,\cdots$ and y_j's are orthonormal in H.

Let P denote the orthogonal projection of H onto the span L of $\{y_1,\cdots\cdots y_n\}$, i.e.

$P\,x = \langle x,y_1\rangle\, y_1 + \cdots\cdots + \langle x,y_n\rangle\, y_n$, $x \in H$.

Let ν be the Gaussian measure in L. Then

$$\tilde{\mu}(T) = \mu\{x \in H;\ (\langle x,y_1\rangle,\cdots, \langle x,y_n\rangle) \in E\}$$

$$= \nu(T \cap L)$$

$$\leq 1 - \nu(C_\varepsilon \cap L) \quad \text{(since } T \cap L \text{ and } C_\varepsilon \cap L \text{ are disjoint).}$$

But $\nu(C_\varepsilon \cap L) \geq \nu\{x \in L;\ ||\ x\ ||_o \leq r\}$

$$= \mu\{x \in H;\ ||P\ x\ ||_o \leq r\}$$

$$= m\{\omega \in \Omega;\ ||P\ x\ ||_o^{\sim}(\omega) \leq r\}$$

$$= 1 - m\{\omega \in \Omega;\ ||P\,x||_o^{\sim}(\omega) > r\}$$

$$\geq 1 - m\{\omega \in \Omega;\ ||\,x\,||_o^{\sim}(\omega) > r\} \text{ (by Lemma 4.3)}$$

$$> 1 - \varepsilon \text{ (by the choice of r).}$$

Therefore,

$$\nu(C_\varepsilon \cap L) \geq 1 - \varepsilon,$$

and

$$\tilde{\mu}(T) < 1 - (1 - \varepsilon) = \varepsilon. \qquad \#$$

In Definition 4.2 and Definition 4.4, we may replace μ by μ_t.

$$\mu_t(E) = \left(\frac{1}{\sqrt{2\pi t}}\right)^n \int_F e^{-\frac{|x|^2}{2t}}\,dx,\ t > 0.$$

μ_t is called the Gauss measure in H of parameter t. Define $\tilde{\mu}_t$ on R_B accordingly. The same proof of Theorem 4.1 works just as well to show that $\tilde{\mu}_t$ has a unique σ-additive extension to the Borel field of B. ($\tilde{\mu} = \tilde{\mu}_1$.)

Notation . p_t will denote the extension of $\tilde{\mu}_t$ to the Borel field $\mathcal{B}$ of B. It is called the Wiener measure with variance t. Later on we will show that $\{p_t\}$ form a contraction semi-group in the Banach space of bounded uniformly continuous functions on B.

Theorem 4.2. The σ-field generated by R_B is the Borel field of B.

Proof. Observe that B is a separable Banach space. As in the case of classical Wiener space (Theorem 3.3) it suffices to show that $\{\|x\| \leq 1\} \in \sigma[R_B]$.

Let $\{a_n; n=1,2,\cdots\}$ be a dense set in B. By the Hahn-Banach theorem, we can pick up $\{z_n; n=1,2,\cdots\} \subset B^*$ such that for all n, $\|z_n\|_{B^*} = 1$ and $(z_n, a_n) = \|a_n\|$.

Then $\{\|x\| \leq 1\} = \bigcap_{n=1}^{\infty} \{x \in B; |(x,z_n)| \leq 1\} \in \sigma[R_B]$.

To see this, let $S = \{\|x\| \leq 1\}$ and $T = \bigcap_{n=1}^{\infty} \{x \in B; |(x,z_n)| \leq 1\}$. Obviously, $S \subset T$. Conversely, suppose $\|x\| = r > 1$. Choose a_n such that $\|x - a_n\| < \frac{r-1}{2}$. Then $\|a_n\| > \frac{r+1}{2}$ and

$$|(x,z_n)| \geq |(a_n,z_n)| - |(x-a_n,z_n)| \geq \|a_n\| - \|x - a_n\| > \frac{r+1}{2} - \frac{r-1}{2} = 1.$$

Hence $x \notin T$. Thus $S \supset T$. #

Theorem 4.3. Suppose a measurable semi-norm $\|\cdot\|$ is given by a symmetric bilinear form $\langle\langle \;,\; \rangle\rangle$ in H, i.e.,

$$\|x\| = \sqrt{\langle\langle x,x \rangle\rangle}, \quad x \in H.$$

Then there exists $T \in L_{(2)}(H)$ such that $\|x\| = |Tx|$, $x \in H$.

Moreover, $\|\cdot\|$ is a norm iff T is injective.

Remark. Compare Exercise 17. See also Exercise 18 and the remark following it.

Proof. Note that the subspace $\{x \in H; \|x\| = 0\}$ of H is closed since $\|x\|$ is continuous by Lemma 4.2. Considering the orthogonal complement of this subspace, if necessary, we may assume that $\|\cdot\|$ is a measurable norm.

Since $\langle\langle x,y\rangle\rangle$ is a continuous bilinear form in H, there exists a bounded linear operator S of H such that
$\langle\langle x,y\rangle\rangle = \langle Sx,y\rangle$.

Obviously, S is self-adjoint, strictly positive definite. Also, by Lemma 4.6, S is a compact operator of H. Let T denote the positive square root of S. Then we have $\|x\| = |Tx|$ for all x in H. By continuity, T extends uniquely to an isometry $\tilde{T}$ from B into H. In fact, $\tilde{T}$ is a unitary operator. To see this, we need only to show that $\tilde{T}$ is onto. Observe that $\tilde{T}(B)$ is a closed subspace of H. Thus it is sufficient to show that the orthogonal complement of $\tilde{T}(B)$ in H is $\{0\}$. Suppose

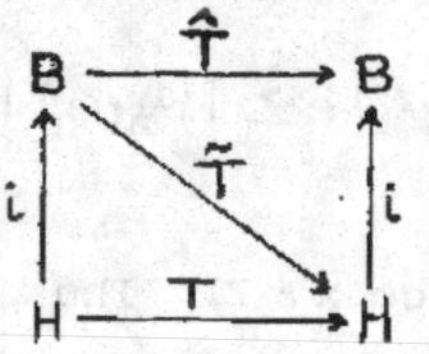

$\langle\tilde{T}x,y\rangle = 0$ for all x in B, where $y \in H$. Then for all z in B

$$\begin{aligned}\langle\langle z,y\rangle\rangle &= \langle\tilde{T}z, \tilde{T}y\rangle \\ &= \langle\tilde{T}z, Ty\rangle \\ &= \langle T\,\tilde{T}\,z, y\rangle \\ &= \langle\hat{T}(\tilde{T}z), y\rangle \\ &= 0.\end{aligned}$$

It follows that the orthogonal complement of $\tilde{T}(B)$ in H is $\{0\}$. Now, consider the Wiener measure p_1 in B. For $x \in B$,

$$\int_B e^{i\langle\langle x,y\rangle\rangle} p_1(dy) = \int_B e^{i\langle \tilde{T}x, \tilde{T}y\rangle} p_1(dy)$$

$$= \int_B e^{i\langle T\tilde{T}x, y\rangle} p_1(dy)$$

$$= e^{-\frac{1}{2}|T\tilde{T}x|^2} \quad \text{(by Lemma 4.7 below)}$$

$$= e^{-\frac{1}{2}\|\tilde{T}x\|^2}.$$

For the sake of clarity, we use $\hat{T}$ to denote i $\tilde{T}$, where i is the inclusion from H into B. Then

$$\int_B e^{i\langle\langle x,y\rangle\rangle} p_1(dy) = e^{-\frac{1}{2}\|\hat{T}x\|^2}$$

Therefore, p_1 is a Gaussian measure in B. Moreover, it is easy to see that $\hat{T} \in L_{(2)}(B)$ by Theorem 2.3.

Finally, let $\{e_i\}$ be an orthonormal basis of B. Then $\{\tilde{T}e_i\}$ is an orthonormal basis of H. But

$$\sum_i |T(\tilde{T}e_i)|^2 = \sum_i \|\tilde{T}e_i\|^2 = \sum_i \|\hat{T}e_i\|^2 < \infty.$$

Hence $T \in L_{(2)}(H)$. #

Recall that B^* is densely embedded in H. Let $z \in B^*, z \neq 0$, then

$$p_t\{x \in B;\ (x,z) < a\}$$

$$= p_t \{x \in B;\ (x, \frac{z}{|z|}) < \frac{a}{|z|}\}$$

$$= \frac{1}{\sqrt{2\pi t}} \int_{-\infty}^{a/|z|} e^{-\frac{u^2}{2t}} du$$

$$= \frac{1}{\sqrt{2\pi t\ |z|^2}} \int_{-\infty}^{a} e^{-\frac{u^2}{2t|z|^2}} du .$$

Hence $z \in B^*$ is normally distributed with mean 0 and variance $t\,|z|^2$ with respect to p_t. Thus we have a densely defined map n from H into $L^2(B,p_t)$. n is an isometry (up to the constant t). Using continuity, we extend n to H. Thus for $h \in H$, $\langle h,\cdot\rangle$ is defined almost everywhere (p_t) in B. Moreover, $\langle h,\cdot\rangle$ is normally distributed with mean 0 and variance $t\,|h|^2$. We have shown the following lemma.

<u>Lemma 4.7.</u> For each $h \in H$, $\langle h,\cdot\rangle$ is defined almost everywhere in B with respect to p_t. It is normally distributed with mean 0 and variance $t\,|h|^2$ in (B, p_t), i.e.,

$$\int_B e^{i\langle h,x\rangle}\ p_t(dx) = e^{-\frac{t}{2}|h|^2} .$$

<u>Exercise 20</u>. Let A be a trace class operator of H. Then $\langle Ax,x\rangle$ is defined almost everywhere in B with respect to p_t and $\int_B \langle Ax,x\rangle\ p_t(dx) = t\,(\text{trace } A)$.

We have seen so far that if we start with a real separable Hilbert space H with Gauss measure μ_t, then we can define measurability of a semi-norm. In general, we have infinitely many choices of measurable norms, e.g., each injective Hilbert-Schmidt operator T gives a measurable norm $\| x \| = |T x|$. For each choice of $\|\cdot\|$, we can complete H with respect to this norm to get a separable Banach space B. In this way, we have (i,H,B). One is led to ask the following question : Does every real separable Banach space arise in this fashion ?

Theorem 4.4. Let B be a real separable Banach space. Then there exists a Hilbert space H densely embedded in B such that B-norm is measurable over H, i.e., (i,H,B) is an abstract Wiener space, where i is the inclusion map from H into B.

Proof. By the separability for B, there exists an increasing sequence of subspaces F_n such that (a) dim $F_n = n$ and (b) the set union $K = \bigcup_{n=1}^{\infty} F_n$ is dense in B. Pick up a sequence $\{z_n;\ n=1,2,\ldots\}$ such that $z_1, z_2, \ldots, z_n$ is a basis of F_n. Let $S = \{x \in B;\ \|x\| < 1\}$.

Step 1. There exists a sequence of positive real numbers $\{\alpha_n\}$ such that if $e_n = \alpha_n z_n$, $n \geq 1$, then

$$\sum_{j=1}^{n} \beta_j e_j \in S \quad \text{whenever} \quad \sum_{j=1}^{n} \beta_j^2 \leq 1.$$

Reason: We choose α_n's inductively. First choose α_1 such that $e_1 = \alpha_1 z_1 \in S$. Having chosen $\alpha_1, \ldots \alpha_{n-1}$, we choose α_n as follows:

Consider $\theta : \mathbb{R}^n \longrightarrow B$

$$(\beta_1,\beta_2,\cdots,\beta_n) \mapsto \sum_{j=1}^{n-1}\beta_j e_j + \beta_n z_n.$$

Clearly θ is continuous. Moreover,

$$\theta^{-1}(S) \supset \{(\beta_1,\beta_2,\ldots,\beta_{n-1},0);\ \sum_{j=1}^{n-1} \beta_j^2 \leq 1\}.$$

Since $\theta^{-1}(S)$ is open and $\{(\beta_1,\beta_2,\ldots\beta_{n-1},0);$ $\sum_{j=1}^{n-1} \beta_j^2 \leq 1\}$ is closed, we can find α_n such that the ellipsoid

$$\{(\beta_1,\beta_2,\ldots,\beta_{n-1},\beta_n);\ \sum_{j=1}^{n-1} \beta_j^2 + (\frac{\beta n}{\alpha_n})^2 \leq 1\}$$

lies between the above two sets. Clearly, we then have the desired property for $e_1,\ldots,e_{n-1}$ and $e_n = \alpha_n z_n$.

Step 2. Define an inner product $< , >_o$ in K by letting $\{e_n\}$ be an orthonormal set. Let $|\cdot|_o = \sqrt{<\cdot,\cdot>_o}$. Then K is a pre-Hilbert space. From step 1, we have, for $x \in K$,

$$|x|_o \leq 1 \implies ||x|| < 1$$

Therefore,

$$||x|| < |x|_o \text{ for all } x \in K.$$

Now, if $\{x_n\}$ is a $|\cdot|_o$-Cauchy sequence in K, then it is also a $||\cdot||$-Cauchy sequence in view of the above relation.

Therefore, $x_n \to x \in B$ since B is complete. Let H_o denote the completion of K in $|\cdot|_o$-norm. We have just seen that $K \subset H_o \subset B$.

Obviously, H_o is dense in B since K is dense in B. But we do not know whether $||\cdot||$ is a measurable norm of H_o or not. To conquer this point, we have to do a little trick in next step.

Step 3. Take an injective Hilbert-Schmidt operator A of H_o and define $|x| = |A^{-1}x|_o$ for $x \in A(H_o)$. Let $H = A(H_o)$. Then H is a Hilbert space in the norm $|\cdot|$. Clearly H is dense in B. Moreover, $||x|| < |x|_o = |Ax|$. The same argument in the proof of Theorem 4.3 shows that $A \in \mathcal{L}_{(2)}(H)$. By Exercise 17, $|Ax|$ is a measurable norm of H. Note that any norm weaker than a measurable norm is also measurable. Hence $||\cdot||$ is also measurable over H. #

Lemma 4.8. Let X be a Banach space and let T be a bounded operator from X into B. If the range of T is contained in H then T is also a bounded operator from X into H.

Proof. Suppose $x_n \to x$ in X and $T x_n \to h$ in H. Since H-topology is stronger than B-topology by Lemma 4.2, $T x_n \to h$ in B. But since T is a continuous linear operator from X into B, we must have $h = T x$. Therefore by the closed graph theorem, T is continuous from X into H. #

Now, suppose T is a bounded operator from B into itself and H is invariant under T, i.e. $T(H) \subset H$. Then, in view of the above lemma, T can be regarded as a bounded operator of H. $||S||_{X,Y}$ will denote the operator norm of S when S is regarded as an operator from X into Y. Thus, it should not be ambiguous in using $||T||_{B,B}$, $||T||_{H,B}$ and $||T||_{H,H}$. Recall that B^* is embedded in B via $B^* \subset H \subset B$. We use $||\cdot||_*$ to denote B^*-norm. It follows from Lemma 4.2 that there exists a constant c such that $|x| \leq c||x||_*$ for all $x \in B^*$.

Theorem 4.5. Let T be a bounded operator of B with finite dimensional range contained in B^*. Then for any $p \geq 1$

$$\int_B ||Tx||^p \, p_t(dx) \leq ||T||_{H,H}^p \int_B ||x||^p \, p_t(dx).$$

Remarks. (a) In general, it is not true that $||Tx|| \leq ||T||_{H,H}||x||$

(b) Later on, we will prove Fernique's theorem:

There exists $\alpha > 0$ such that $\int_B e^{\alpha||x||^2} p_t(dx) < \infty$.

Therefore, the quantity $\int_B ||x||^p \, p_t(dx)$ is finite for all $p \geq 1$.

Proof. Let M be the range of T and μ_t be the Gaussian measure in M with variance t. Let P be the orthogonal projection of H onto M. Define

$\theta(\lambda) = p_t\{x \in B; \ ||x|| > \lambda\}$

$\xi(\lambda) = \mu_t\{x \in M; \ ||x|| > \lambda\}$

and $\eta(\lambda) = \mu_t\{x \in M; ||Tx|| > \lambda\}$.

Note that $\xi(\lambda) = p_t\{x \in B; ||Px|| > \lambda\}$.

It follows from Lemma 4.3 that

(a) $\eta(\lambda) \leq \xi(\lambda/||T||_{H,H})$ (with M replacing B)

and (b) $\xi(\lambda) \leq \theta(\lambda)$.

Therefore, we have

$$\int_B ||Tx||^p \, p_t\,(dx) = \int_M ||Tx||^p \, \mu_t\,(dx)$$

$$= -\int_0^\infty \lambda^p \, d\eta\,(\lambda)$$

$$= p \int_0^\infty \eta(\lambda)\,\lambda^{p-1}\, d\lambda$$

$$\leq p \int_0^\infty \xi\left(\frac{\lambda}{||T||_{H,H}}\right)\lambda^{p-1}\, d\lambda$$

$$= ||T||_{H,H}^p \, p \int_0^\infty \xi(\tau)\, \tau^{p-1}\, d\tau$$

$$\leq ||T||_{H,H}^p \, p \int_0^\infty \theta(\lambda)\, \lambda^{p-1}\, d\lambda$$

$$= ||T||_{H,H}^p \int_B ||x||^p \, p_t\,(dx). \qquad \#$$

<u>Theorem 4.6.</u> (Goodman). Let A be a bounded linear operator of B with range in B*. Then A is a trace class operator of H. Moreover,

$$||A||_1 \leq \int_B ||x||^2 \, p_1\,(dx)\cdot ||A||_{B,B^*}.$$

Proof. Let T be an operator as in Theorem 4.5. By Exercise 20, trace $TA = \int_B (TAx,x)\, p_1\,(dx)$.
Hence,

$$|\text{trace } TA| \leq \int_B |(TAx,x)| p_1(dx)$$

$$= \int_B |\langle Ax, T^*x\rangle| \, p_1(dx)$$

$$\leq \int_B ||Ax||_* ||T^*x|| \, p_1(dx)$$

$$\leq ||A||_{B,B^*} \int_B ||x|| \; ||T^*x|| \, p_1(dx)$$

$$\leq ||A||_{B,B^*} \sqrt{\int_B ||x||^2 p_1(dx)} \sqrt{\int_B ||T^*x||^2 p_1(dx)}$$

$$\leq ||A||_{B,B^*} \sqrt{\int_B ||x||^2 p_1(dx)} \sqrt{||T^*||^2_{H,H} \int_B ||x||^2 p_1(dx)}$$

(by Theorem 4.5)

$$= ||A||_{B,B^*} \int_B ||x||^2 p_1(dx) \cdot ||T^*||_{H,H}$$

$$= \int_B ||x||^2 p_1(dx) \cdot ||A||_{B,B^*} ||T||_{H,H}.$$

The above relation holds for all operators T as in Theorem 4.5. It is easy to see that it holds for all degenerate operator T of H. Hence by Remark (c) following Theorem 1.7, it holds for all $T \in \mathcal{X}(H)$.

That is, we have for all $T \in \mathcal{X}(H)$

$$|\text{trace } TA| \leq \int_B ||x||^2 \, p_1(dx) \cdot ||A||_{B,B^*} ||T||_{H,H}.$$

But $\mathcal{X}(H)^* = \mathcal{L}_{(1)}(H)$ and the pairing between $\mathcal{X}(H)$ and $\mathcal{L}_{(1)}(H)$ is given by $(E,F) = \text{trace } EF$. Therefore, we have shown not only hat $A \in \mathcal{L}_{(1)}(H)$, but also $||A||_1 \leq \int_B ||x||^2 \, p_1(dx) \cdot ||A||_{B,B^*}$. #

orollary 4.4. (Kuo) Let C be a bounded operator of B with range n H. Then C is a Hilbert-Schmidt operator of H. oreover,

$$||C||_2 \leq \{\int_B ||x||^2 \, p_1(dx)\}^{1/2} ||C||_{B,H}.$$

roof. Let $C|_H$ denote the restriction of C to H. We show that $(C|_H)^*$ has range in B^*. Let $x_o \in H$ then

$$\sup_{\substack{x \in H \\ x \neq 0}} |\langle (C|_H)^* x_o, x\rangle| / ||x|| = \sup_{\substack{x \in H \\ x \neq 0}} |\langle x_o, C x\rangle| / ||x|| \leq |x_o| \cdot ||C||_{B,H}.$$

nce $(C|_H)^* x_o \in B^*$ by Exercise 19 and $||(C|_H)^* x_o||_* \leq ||C||_{B,H} |x_o|$. erefore,

$$||(C|_H)^*||_{H,B^*} \leq ||C||_{B,H}.$$

Let $A = (C|_H)^* C$, then A is a bounded operator from B into B^*. Theorem 4.6, $A \in \mathcal{L}_{(1)}(H)$ and

$$||A||_1 \leq \int_B ||x||^2 \, p_1(dx) ||A||_{B,B^*}.$$

refore $C \in \mathcal{L}_{(2)}(H)$. Moreover, note that

$$||A||_1 = \text{trace } (C|_H)^* C = ||C||_2^2$$

and $||A||_{B,B^*} \leq ||(C|_H)^*||_{H,B^*}||C||_{B,H} \leq ||C||^2_{B,H}$.

Hence, $||C||_2 \leq \{\int_B ||x||^2 p_1(dx)\}^{1/2}||C||_{B,H}$. #

§5. C[0,1] as an abstract Wiener space.

Let us recall the definition of an abstract Wiener space. Take a real separable Hilbert space H with norm $|\cdot|$. Take a measurable norm $||\cdot||$ on H (see Definition 4.4). By Lemma 4.2 $||\cdot||$ is weaker than $|\cdot|$. Let B be the completion of H with respect to $||\cdot||$. Let i be the inclusion map from H into B. The triple (i,H,B) is called an abstract Wiener space. Note that i is a compact operator by Lemma 4.6. One example of such triples is where B is given by a one-to-one Hilbert-Schmidt operator of H (see Exercise 17). Actually, by Theorem 4.3 we see that if B is a Hilbert space then it is given by a Hilbert-Schmidt operator of H.

Exercise 21. Show that $(i, L^2[0,1], L^p[0,1])$ is not an abstract Wiener space, where $1 \leq p < 2$ and i is the inclusion map from $L^2[0,1]$ into $L^p[0,1]$.

Exercise 22. Show that $(i, \mathcal{L}_{(2)}(H), \mathcal{K}(H))$ is not an abstract Wiener space.

As a matter of fact, the notion of an abstract Wiener space stems from C[0,1]. We will define a subspace C' of C[0,1] below. C' plays an important role in the integration theory in C[0,1]. C' is a Hilbert space with its norm and (i,C', C[0,1]) is an abstract Wiener space. In fact, this is the reason (i, H,B) gets its name. We will call (i, C', C[0,1]) the classical Wiener space.

Before defining C', we introduce stochastic integral (sometime, it is also called Wiener integral, but we reserve Wiener integral for the integral associated with the Wiener measure). Let f be a simple function on [0,1] of the form,

$$f = \sum_{j=1}^{n} a_j 1_{[t_j, t_{j+1})},$$

where $0 \leq t_1 < t_2 < \cdots < t_{n+1} \leq 1$.

Define $\Theta(f): C[0,1] \rightarrow \mathbb{R}$ by

$$\Theta(f)(x) = \sum_{j=1}^{n} a_j (x(t_{j+1}) - x(t_j)).$$

Note that $x(t_{j+1})-x(t_j)$, $j=1,2,\ldots,n$, are independent Gaussian variables in C[0,1]. Moreover, $x(t_{j+1})-x(t_j)$ has mean 0 and variance $t_{j+1}-t_j$. Therefore, $\Theta(f)$ is also a Gaussian variable with mean 0 and variance

$$\sum_{j=1}^{n} a_j^2 (t_{j+1}-t_j) = \int_0^1 f(t)^2 \, dt.$$

Thus the map

$$f \longmapsto \theta(f)$$

is a densely defined isometry from $L^2[0,1]$ into $L^2(C[0,1],w)$. This map extends uniquely to $L^2[0,1]$.

Notation. $\Theta(f)(x) \equiv \int_0^1 f(t)dx(t)$. It is called a stochastic integral.

Remark. For general $f \in L^2[0,1]$, $\int_0^1 f(t)dx(t)$ is defined w-a.e. in $C[0,1]$.

Theorem 5.1. $\int_0^1 f(t)dx(t)$ is a Gaussian variable with mean 0 and variance $\int_0^1 f(t)^2 dt$.

Proof. Choose a sequence of simple functions f_j converging to f in $L^2[0,1]$. Then $\Theta(f_j)$ converges to $\Theta(f)$ in $L^2(C[0,1],w)$. Choose a subsequence, still denoted by $\Theta(f_j)$, such that $\Theta(f_j)$ converges to $\Theta(f)$ w-a.e. Then

$$\int_C e^{i\Theta(f)(x)}\, w(dx) = \lim_{j\to\infty} \int_C e^{i\Theta(f_j)(x)}\, w(dx)$$

$$= \lim_{j\to\infty} e^{-\frac{1}{2}\int_0^1 f_j(t)^2\, dt}$$

$$= e^{-\frac{1}{2}\int_0^1 f(t)^2 dt}.$$

Hence $\Theta(f)$ is normally distributed with mean 0 and variance $\int_0^1 f(t)^2 dt$. #

Theorem 5.2. $E_w[\int_0^1 f(t)dx(t) \int_0^1 g(t)dx(t)] = \int_0^1 f(t)g(t)dt$.

Proof. Since $\Theta: L^2[0,1] \to L^2(C[0,1],w)$ is an isometry. #

Let C' consist of all $\mathbb{R}$-valued functions f such that f is absolutely continuous, $f(0) = 0$ and $f' \in L^2[0,1]$. Define an inner product $\langle\ ,\ \rangle$ in C' by

$$\langle f,g\rangle = \int_0^1 f'(t)\, g'(t)dt.$$

Then C' is a Hilbert space and $C' \subset C$. In fact, if we define an operator D from $L^2[0,1]$ to C' by

$$Df(t) = \int_0^t f(s)ds, \quad f \in L^2[0,1],$$

then D is a unitary operator from $L^2[0,1]$ onto C'. Note that

$$D^*f(t) = f'(t), \quad f \in C'.$$

Let $|\cdot|$ denote the norm of C'. Obviously, $\|x\| = \sup_{0\le t\le 1} |x(t)|$ is a weaker norm than $|\cdot|$ in C' because if $f \in C'$ then for all $0 \le t \le 1$,

$$|f(t)| = \left|\int_0^t f'(s)ds\right| \le \int_0^t |f'(s)|ds \le \int_0^1 |f'(s)|ds \le |f|.$$

Observe that C' is separable because it is unitary equivalent to $L^2[0,1]$. Let i be the inclusion map of C' into C[0,1].

Theorem 5.3. (i, C', C[0,1]) is an abstract Wiener space.

Remark. Assuming this theorem, we connect stochastic integral and the inner product of C'. Let $f \in C'$ be fixed, then

$$\langle f,g\rangle = \int_0^1 f'(t)g'(t)dt = \int_0^1 f'(t)dg(t), \quad g \in C'.$$

By Lemma 4.7, $\langle f,g\rangle$ extends to C[0,1] as a Gaussian random variable with mean 0 and variance $\int_0^1 f'(t)^2 dt$. Therefore, we have for $f \in L^2[0,1]$,

$$\int_0^1 f(t)dx(t) = \langle Df, x \rangle, \quad x \in C[0,1] \text{ (w-a.e.)}.$$

This relation can be used to give another proof of Kac's formula (Theorem 3.4). The proof is left as Exercise 23.

The proof of Theorem 5.3 follows from the three lemmas below.

Lemma 5.1. $C[0,1]$ is the completion of C' with respect to the sup norm.

Proof. Let A_n consist of those functions x in $C[0,1]$ such that x takes rational values at $\frac{k}{2^n}$, $k=1,2,\cdots,2^n$, and is linear between these binary points. Let $A = \bigcup_{n=1}^{\infty} A_n$. It is easy to see that A is dense in $C[0,1]$. Therefore, C' is dense in $C[0,1]$ because $A \subset C'$. #

Lemma 5.2. For each $\varepsilon > 0$, $w\{x \in C[0,1];\ ||x|| \leq \varepsilon\} > 0$.

Proof. For each $x \in C[0,1]$, let w_x be the measure defined by

$$w_x(E) = w(E-x), \quad E \in \mathcal{B}(C[0,1]).$$

We will prove in the next chapter that w_x is equivalent to w iff $x \in C'$.

From the proof of Lemma 5.1, we can pick up a countable dense subset $\{z_1, z_2, \ldots, z_n, \ldots\}$ of $C[0,1]$ such that $z_n \in C'$ for all n.

Let $b(x,r) \equiv \{y \in C[0,1];\ ||y-x|| \leq r\}$, then for any $\varepsilon > 0$

$$C[0,1] = \bigcup_{n=1}^{\infty} b(z_n, \varepsilon).$$

Now, suppose for some $\varepsilon > 0$, $w\{x \in C[0,1]; ||x|| \leq \varepsilon\} = 0$. Then $w\{b(z_n, \varepsilon)\} = w_{z_n}\{x \in C[0,1]; ||x|| \leq \varepsilon\} = 0$ for all n. Therefore, $w\{C[0,1]\} = 0$, which yields the desired contradiction. #

Lemma 5.3. $||\cdot||$ is a measurable norm over C'.

Proof. For each n, we define a semi-norm $||\cdot||_n$ in C' by

$$||x||_n = \sup\{|x(t_j)|;\ t_j = \frac{j}{2^n},\ j=1,2,\ldots 2^n\}.$$

Clearly, $||\cdot||_n$ is a measurable semi-norm in C' and $\lim_{n\to\infty} ||x||_n = ||x||$ for each x in C'. Conditions (1) and (2) of [16, Theorem 4] are obviously satisfied, while condition (3) follows from Lemma 5.2. Hence by that theorem $||\cdot||$ is measurable in C'. #

Another proof of Lemma 5.3 can be done by using Dudley-Feldman-LeCam's theorem [9], Theorem 3.1 and Lemma 5.4 below. Let $(H, |\cdot|)$ be a real separable Hilbert space with Gauss measure μ (see Definition 4.2). Let $||\cdot||$ be a norm in H weaker than $|\cdot|$ and B the completion of H with respect to $||\cdot||$. As before, μ induces $\tilde{\mu}$ defined in the cylinder subsets of B. Dudley, Feldman and LeCam have shown that $\tilde{\mu}$ has σ-additive extension to the Borel field of B if and only if $||\cdot||$ is a measurable norm in H. Therefore, in view of Theorem 3.1, to show Lemma 5.3 it is sufficient to prove the following lemma.

Lemma 5.4. w is the σ-additive extension of the Gauss measure in C' to the Borel field of C[0,1].

Proof. By Theorem 5.1, Theorem 5.2 and the unitary operator D from $L^2[0,1]$ onto C' defined before. #

§6. Weak distribution and Gross-Sazonov theorem.

Motivated by the study of quantum field theory, Segal [41] makes the following

Definition 6.1. A weak distribution on a topological vector space L is an equivalence class of linear maps F from the topological dual space L* to the random variables on a probability space (depending on F). Two such maps F_1 and F_2 are equivalent if for every finite $y_1, y_2, \ldots y_k$ in L*,

$$\{F_1(y_1), F_1(y_2), \ldots, F_1(y_k)\} \text{ and } \{F_2(y_1), F_2(y_2), \ldots, F_2(y_k)\}$$

have the same joint distribution.

The notion of weak distributions is equivalent to that of cylinder set measures. Let L be a topological vector space. A cylinder set in L is a set of the form $\{x \in L; (y_1(x), \ldots, y_k(x)) \in E\}$, where $y_1, \ldots, y_k \in L^*$ and $E \in \mathcal{B}(\mathbb{R}^k)$. If K is a finite dimensional subspace of L* containing $y_1, \ldots, y_k$ then we say that this cylinder set is based on K. Let $\mathcal{R}$ be the collection of cylinder sets and $\mathcal{R}_K$ those based on K. Clearly, $\mathcal{R}$ is a field and $\mathcal{R}_K$ is a σ-field.

Definition 6.2. A cylinder set measure on L is a nonnegative finitely additive function ν defined on $\mathcal{R}$ such that $\nu(L) = 1$ and ν is σ-additive on $\mathcal{R}_K$ for any finite dimensional subspace K of L*.

Exercise 24. Show that there is a one-to-one correspondence between weak distributions and cylinder set measures such that if F corresponds to ν then

$$\text{Prob}\ \{(F(y_1),\ldots,F(y_k))\in E\} = \nu\{x; (y_1(x),\ldots,y_k(x))\in E\}.$$

Example 1. Let H be a real separable Hilbert space. Let n be the linear map given in Corollary 4.1. The weak distribution represented by n is called the normal distribution on H with parameter 1.

Example 2. The weak distribution corresponding to the cylinder set measure μ_t defined in §4 is called the normal distribution on H with parameter t. It is denoted by n_t. For h in H, $n_t(h)$ is normally distributed with mean 0 and variance $t\cdot|h|^2$.

Given a weak distribution represented by F on a topological vector space L, we can build an integration theory in L. Specifically, let f be a cylinder function on L, i.e., a function of the form

$$f(x) = \phi(y_1(x),\ldots,y_k(x)),$$

where ϕ is a Borel function in $\mathbb{R}^k$ and $y_1,\ldots,y_k \in L^*$. Define a random variable

$$F(f) = \phi(F(y_1),\ldots,F(y_k)),$$

and if F(f) is integrable then we define

$$\int f = E(F(f)), \quad \text{(E: expectation)}.$$

It is easy to see that $\int f$ is independent of the particular representation F. However, in order for this integration to be useful, we must be able to integrate a wider class of functions. In the case of normal distribution, we have a satisfactory result.

A special class of such functions is given in Theorem 6.2. For a more general discussion, see [16]. Another question is whether we can pick up a nice representation. Theorem 6.1 says that an abstract Wiener space provides such a representation.

<u>Theorem 6.1</u>. Let H be a real separable Hilbert space. Let $||\cdot||$ be a measurable norm in H and B the completion of H with respect to $||\cdot||$ so that (i,H,B) is an abstract Wiener space. Let p_t be the Wiener measure of B with parameter t. Then the identity map on B* regarded as densely defined from H* into the random variables over the probability space (B,p_t) extends to a representation of n_t.

Proof. This follows easily from Lemma 4.7. #

Let $\mathcal{T}_m$ denote the weakest topology on H on which all measurable semi-norms are continuous. Recall that if $A \in \mathcal{L}_{(2)}(H)$ then $|Ax|$ is a measurable semi-norm.

<u>Exercise 25</u>. $E[|Ax|^{\sim}]^2 = ||A||_2^2$, $A \in \mathcal{L}_{(2)}(H)$.

<u>Theorem 6.2</u>. Let f be a complex-valued function in H satisfying the condition: there exist two measurable semi-norms $||\cdot||_o$ and $||\cdot||$ such that f is uniformly continuous in $\{x \in H;\ ||x||_o \leq r\}$ for each $r > 0$ with respect to $||\cdot||$. Then the net $(f\circ P)^{\sim}$ of random varialbes converges in probability as $P \to I$ strongly through $\mathcal{F}$. (The limit will be denoted by $\tilde{f}$).

Remark. Obviously, if f is uniformly $\mathcal{T}_m$-continuous in H then f satisfies the above condition.

Proof. Recall that $||Px||_o^{\sim}$ converges to $||x||_o^{\sim}$ in probability as $P \to I$ through $\mathcal{F}$ (Lemma 4.1) and $||x||_o^{\sim}$ is integrable (see

Remark (b) following Theorem 4.5). Therefore, given $\varepsilon > 0$, there exist $r > 0$ and $P' \in \mathcal{F}$ such that

$\text{prob}\,\{||Px||_o^\sim > r\} < \varepsilon/3$ whenever $P \in \mathcal{F}$ and $P \geq P'$.

By the assumption on f, there exists $\delta > 0$ such that whenever $||x||_o \leq r$, $||y||_o \leq r$ and $||x - y|| < \delta$, we have

$$|f(x) - f(y)| < \varepsilon$$

On the other hand, it follows from the definition of measurability for $||\cdot||$ that there exists $P'' \in \mathcal{F}$ such that $\text{prob}\,\{||Px||^\sim \geq \delta\} < \varepsilon/3$ whenever $P \in \mathcal{F}$ and $P \perp P''$.

Take $P_o \in \mathcal{F}$ such that $P_o > P'$ and $P_o > P''$. Then, whenever $P, Q > P_o$, we have

$\{x \in H; ||Px||_o \leq r, ||Qx||_o \leq r, \text{ and } ||Px-Qx|| < \delta\}$

$\subseteq \{x \in H;\ |f(Px)-f(Qx)| < \varepsilon\}$.

Hence, if $P,Q > P_o$,

$\text{prob}\,\{\,|f(Px)^\sim - f(Qx)^\sim| \geq \varepsilon\}$

$\leq \text{prob}\,\{||Px||_o^\sim > r\} + \text{prob}\,\{||Qx||_o^\sim > r\} + \text{prob}\,\{||(P-Q)x||^\sim \geq \delta\}$

$< \varepsilon/3 + \varepsilon/3 + \varepsilon/3$

$= \varepsilon$.

Now, let $\varepsilon = 1/n$ and denote by P_n the orthogonal projection P_o obtained as above. We may assume that P_n is increasing. It is easy to see that $(f \circ P_n)^\sim$ converges in probability to $\tilde{f}$, say. The same argument as in the proof of Lemma 4.1 shows that in fact the net $(f\circ P)^{\sim}$ converges in probability to $\tilde{f}$ as $P \to I$ through $\mathcal{F}$. #

Exercise 26. Let f satisfy the condition in Theorem 6.2. Show that $f \equiv 0$ if and only if $\tilde{f} = 0$ almost surely.

Theorem 6.3. Let (i, H,B) be an abstract Wiener space. Let g be a continuous function in B. Let f be the restriction of g to H, i.e., $f = g\circ i$. Then f satisfies the condition in Theorem 6.2 and $\tilde{f} = g$ almost everywhere (w.r.t. p_1).

Remark. We use Theorem 6.1 to regard (B,p_1) as a representing probability space for the normal distribution n_1 so that $\tilde{f}$ is defined in B.

Proof. We can take a sequence $\{a_n\}$ satisfying both conditions in the proofs of Corollary 4.2 and Lemma 4.5, e.g., $a_n = n^2+1$, so that we have the conclusions of Corollary 4.2 and Lemma 4.5 simultaneously. $||\cdot||_o$ denotes B_o-norm and the same notation used there will be adopted below.

Since for each $r > 0$, $\{x \in H;\ ||x||_o \leq r\}$ is precompact with respect to $||\cdot||$ and g is $||\cdot||$-continuous in B, f satisfies the condition in Theorem 6.2.

Observe that the Gauss measure μ_1 has σ-additive extensions to $\mathcal{B}(B_o)$ and $\mathcal{B}(B)$. Therefore, $p_1(B_o) = 1$. Take the sequence $\{P_n\}$ and its associated sequence $\{P_n^{\sim}\}$ in Corollary 4.2. Obviously,

$$\tilde{f} = \lim_{n\to\infty} \text{in prob } (f\circ P_n)^{\sim} .$$

But, if $x \in B_o$, we have

$$(f \circ P_n)^{\sim}(x) = g(\tilde{P}_n x)$$
$$\to g(x) \text{ as } n \to \infty ,$$

by Corollary 4.2 (a),(c) and by the continuity of g.

Therefore $\tilde{f}(x) = g(x)$ for all $x \in B_o$. That is, $\tilde{f} = g$ almost everywhere w.r.t. p_1. #

In order to prove Gross-Sazonov's theroem, we state Prohorov's theorem [38] without proof.

Notation. $\mathcal{M}_X$ = the space of positive finite Borel measures in a separable complete metric space (X,ρ).

Definition 6.3. A sequence $\{\mu_n\}$ in $\mathcal{M}_X$ is said to converge weakly to μ in $\mathcal{M}_X$ if $\lim_{n\to\infty} \int_X f\, d\mu_n = \int_X f\, d\mu$ for every bounded continuous function f in X.

Definition 6.4. The Prohorov metric d of μ and ν on $\mathcal{M}_X$ is defined to be the infimum of $\varepsilon > 0$ such that

$$\mu(F) \leq \nu(F^{\varepsilon}) + \varepsilon \text{ and } \nu(F) \leq \mu(F^{\varepsilon}) + \varepsilon$$

for every closed subset F of X, where $F^{\varepsilon} = \{x \in X;\ \rho(x,F) < \varepsilon\}$.

Theorem 6.4 (Prohorov).

(a) $\mathcal{M}_X$ is a separable complete metric space with the Prohorov metric.

(b) The topology induced by the Prohorov metric is the same as that induced by the weak convergence.

(c) A subset $\mathcal{N} \subset \mathcal{M}_X$ is precompact if and only if (i) there exists $M < \infty$ such that $\mu(X) \leq M$ for all $\mu \in \mathcal{N}$ and (ii) for every $\varepsilon > 0$, there exists a compact subset K_{ε} of X such that

$$\mu(K_{\varepsilon}^{c}) < \varepsilon \quad \text{for all } \mu \in \mathcal{N}.$$

From now on, we will fix a real separable Hilbert space H and $\mathcal{M}$ will denote $\mathcal{M}_H$. Let $\mu \in \mathcal{M}$. It follows from Theorem 2.2 that the characteristic functional $\hat{\mu}$ of μ is uniformly $\mathcal{T}_m$-continuous. Therefore, by Theorem 6.2 and the remark following it, $(\hat{\mu})^\sim$ makes sense as a random variable. In this way, we can associate a random variable with every μ in $\mathcal{M}$.

<u>Theorem 6.5</u>. Let a sequence $\{\mu_n\}$ in $\mathcal{M}$ converge weakly to μ. Let $\phi_n = \hat{\mu}_n$ and $\phi = \hat{\mu}$. Then $\phi_n^\sim$ converges to $\phi^\sim$ in probability.

Proof. Actually, we can prove a stronger conclusion, namely, $E|\phi_n^\sim - \phi^\sim|^2 \to 0$ as $n \to \infty$. Clearly,

$$E|\phi_n^\sim - \phi^\sim|^2 = E\int_H \int_H \{e^{i<\cdot\, ,x-y>}\}^\sim (\mu_n(dx)-\mu(dx))(\mu_n(dy)-\mu(dy)),$$

where $\{e^{i<\cdot\, ,x-y>}\}^\sim$ denotes the random variable corresponding to the function $e^{i<h,x-y>}$ of $h \in H$.

Note that $E\{e^{i<\cdot\, ,x-y>}\}^\sim = e^{-\frac{1}{2}|x-y|^2}$. Therefore,

$$E|\phi_n^\sim - \phi^\sim|^2 = \int_H \int_H e^{-\frac{1}{2}|x-y|^2} (\mu_n(dx)-\mu(dx))(\mu_n(dy)-\mu(dy)).$$

Suppose $\varepsilon > 0$. It follows from the assumption and Theorem 6.4 that there exists a compact set K such that

$$\mu_n(K^c) < \varepsilon/4 \text{ and } \mu(K^c) < \varepsilon/4.$$

Writing the last double integral as $\int_K \int_H + \int_{K^c} \int_H$, we see that $|\int_{K^c}\int_H| < \varepsilon$. So we need to show that $\lim_{n\to\infty} |\int_K \int_H| = 0$ for every compact subset K of H.

Now, let $\delta > 0$ be given. Then there is a compact set L such that

$$\mu_n(L^c) < \delta \text{ and } \mu(L^c) < \delta .$$

Since $L \times K$ is compact in $X \times X$, there exists an open covering $U(y_1), U(y_2), \ldots U(y_m)$ of K such that for all $x \in L$,

$$|f(x,y) - f(x,y_j)| < \delta \quad , \text{ whenever } y \in U(y_j),$$

where $f(x,y) = e^{-\frac{1}{2}|x-y|^2}$.

But $\left|\int_H f(x,y)(\mu_n(dx) - (dx))\right|$

$$\leq \left|\int_L f(x,y)(\mu_n(dx) - \mu(dx)\right| + 2\delta$$

$$\leq \left|\int_L f(x,y_j)(\mu_n(dx) - \mu(dx)\right| + 4\delta$$

$$\leq \left|\int_H f(x,y_j)(\mu_n(dx) - \mu(dx))\right| + 6\,\delta, \; j = 1,2,\ldots,m.$$

Therefore, $\left|\int_H f(x,y)(\mu_n(dx) - \mu(dx))\right| \leq 7\,\delta$ for large n.

Hence

$$\left|\int_K \int_H\right| \leq \int_K \left|\int_H f(x,y)(\mu_n(dx) - \mu(dx))\right| (\mu_n(dy) + \mu(dy))$$

#

Theorem 6.6. Let $\{\mu_n\}$ be a sequence in $\mathcal{M}$. Let $\phi_n = \hat{\mu}_n$. Assume ϕ is a function in H such that $\phi(0) = 1$ and ϕ is uniformly continuous in $\mathcal{T}_m$-topology. Assume also that $\phi_n^\sim$ converges to $\phi^\sim$

in probability. Then there exists $\mu \in \mathcal{M}$ such that $\hat{\mu} = \phi$ and μ_n converges to μ weakly.

Proof. Suppose we have $\mu \in \mathcal{M}$ such that μ_n converges to μ weakly. Let $\psi = \hat{\mu}$. Then by Theorem 6.5,

$$\psi^{\sim} = \lim_{n\to\infty} \text{ in prob } \phi_n^{\sim}.$$

Thus $\psi^{\sim} = \phi^{\sim}$. Hence $(\psi - \phi)^{\sim} = 0$ and by Exercise 26 $\psi - \phi \equiv 0$. That is $\phi = \hat{\mu}$.

Thus we need only to show that there exists $\mu \in \mathcal{M}$ such that μ_n converges to μ weakly. By Theorem 6.4, it suffices to show that for any $\varepsilon > 0$, there is a compact set K in H such that $\mu_n(K^c) < \varepsilon$, $n \geq 1$. In fact, we will show that there exists a precompact set S such that $\mu_n(S) > 1 - \varepsilon$ for all large n.

Step 1: We show that $|\phi(x)| \leq 1$ for all $x \in H$.

Let f be a non-negative smooth function in $\mathbb{C}$, having compact support and vanishing in the unit disk. Then $f(\phi(\cdot))$ is also uniformly continuous in $\mathcal{T}_m$-topology, so are $f(\phi_n(\cdot))$, $n \geq 1$. But $f(\phi(\cdot))^{\sim} = \lim_{n\to\infty}$ in prob $f(\phi_n(\cdot))^{\sim} = 0$. Hence $f(\phi(x)) = 0$ for all x. Hence $|\phi(x)| \leq 1$ for all x.

Step 2: Let $\delta = \delta(\varepsilon) > 0$ (to be specified later). Then there exists a measurable semi-norm $||\cdot||$ such that

$$||x|| < 1 \Longrightarrow |\phi(x) - 1| < \delta .$$

Exercise 27. Let $||\cdot||_2$ be a measurable semi-norm. Show that there exists a measurable norm $||\cdot||_3$ such that $||x||_2 \leq ||x||_3$ for all x in H.

Therefore, without loss of generality, we can and will assume that $||\cdot||$ is a measurable norm. It is easy to see that the above relation implies that

$$\text{Re } \phi(x) > 1-\delta-2 \, ||x||^2, \; x \in H.$$

Step 3: By Lemma 4.6 and its proof, there exist an injective, self-adjoint compact operator C and another measurable norm $||\cdot||_o$ such that $||x|| \leq ||C\, x||_o$. Therefore, we have

$$\text{Re } \phi(x) > 1-\delta-2 \, ||C\, x||_o^2, \; x \in H.$$

Now, let B and B_o be the completion of H with respect to $||\cdot||$ and $||\cdot||_o$, respectively. Then by continuity C^{-1} extends to a bounded operator from B_o into B. Since ϕ is assumed to be uniformly continuous in $\mathcal{T}_m$-topology, $\tilde{\phi}$ makes sense as a random variable defined in B by Theorem 6.1 and Theorem 6.2. Thus the above inequality extends to B_o as follows:

$$\text{Re } \tilde{\phi}(C^{-1}y) > 1-\delta-2 \, ||y||_o^2, \; y \in B_o.$$

Step 4: Let E_o denote the expectation with respect to the Wiener measure p_1 in B_o. Let $a = 2\, E_o||y||_o^2$. Then upon taking expectation, we have

$$\text{Re } E_o[\, \tilde{\phi}(C^{-1}y)] > 1-\delta-a.$$

But $\tilde{\phi}_n$ converges to $\tilde{\phi}$ in probability by assumption and $|\tilde{\phi}_n| \leq |$, $|\tilde{\phi}| \leq 1$ by step 1. Hence, by taking a subsequence, if necessary, we have $E_o[\tilde{\phi}_n(C^{-1}y)] \to E_o[\,\tilde{\phi}(C^{-1}y)]$ as $n \to \infty$. Therefore, for large n,

$$\text{Re } E_o[\tilde{\phi}_n(C^{-1}y)] > 1-\delta-a.$$

Step 5: It is easy to see that

$$E_o[\tilde{\phi}_n(C^{-1}y)] = \int_{|C^{-1}x| < \infty} e^{-|C^{-1}x|^2/2} \mu_n(dx).$$

Let $S = \{x \in H;\; |C^{-1}\, x| \leq \sqrt{2}\}$. Then

$$E_o[\tilde{\phi}_n(C^{-1}y)] \leq \mu_n(S) + \frac{1}{e}\mu_n(S^c)$$

$$= (1 - \frac{1}{e})\mu_n(S) + \frac{1}{e}.$$

Therefore, from step 4, we have

$$(1 - \frac{1}{e})\ \mu_n(S) + \frac{1}{e} > 1-\delta-a,$$

i.e.,

$$\mu_n(S) > \ 1 - \frac{\delta+a}{1-\frac{1}{e}}\ .$$

Finally, by multiplying a constant to $||\cdot||_o$, we can make a as small as we want (of course, the compact operator C would be affected by a constant multiple). So we choose $||\cdot||_o$ such that

$$a < \varepsilon\ (1 - \frac{1}{e})$$

Then we choose $\delta < 0$ such that

$$\delta < \varepsilon(1 - \frac{1}{e})-a.$$

Hence

$$\mu_n(S) > \ 1-\varepsilon \quad \text{for all large n.}$$

To finish the proof, simply observe that $S = C\ \{x\in H;\ |x| \leq \sqrt{2}\}$ is precompact because C is a compact operator. #

<u>Theorem 6.7</u> (Gross-Sazonov) A functional ϕ in H is the characteristic functional of a probability measure in H if and only if ϕ is $\mathcal{T}_m$-continuous, positive definite and $\phi(0) = 1$.

Proof. Necessity was proved before, (see the remark above Theorem 6.5). To show the sufficiency, let P_n be a sequence in $\mathfrak{F}$ convergin strongly to identity. Let $\phi_n = \phi(P_n)$. By the finite dimensional Bochner's theorem, there eisits μ_n such that $\phi_n = \hat{\mu}_n$. Clearly $\phi_n^{\sim}$

converges to $\phi^{\sim}$ in probability by Theorem 6.2. Therefore, by Theorem 6.6, there exists $\mu \in \mathcal{M}$ such that $\phi = \hat{\mu}$. #

§7. Comments on Chapter I.

1. In [33] Hilbert-Schmidt type n-linear maps in a separable Hilbert space H are defined in a similar way as in Definition 1.1. Specifically, let $\mathcal{L}^n(H)$ denote the continuous n-linear maps $T: \underbrace{H \times H \times \cdots \times H}_{n} \to \mathbb{R}$. T is said to be of Hilbert-Schmidt type if

$$\sum_{i_1,\ldots,i_n=1}^{\infty} T(e_{i_1},\cdots,e_{i_n})^2$$

is convergent for any orthonormal basis $\{e_k\}_{k=1}^{\infty}$ of H. Define

$$||T||_2 = \{\sum_{i_1,\ldots,i_n=1}^{\infty} T(e_{i_1},\ldots,e_{i_n})^2\}^{1/2}.$$

Let $\mathcal{L}^n_{(2)}(H)$ denote the collection of Hilbert-Schmidt type n-linear maps. Then $\mathcal{L}^n_{(2)}(H)$ is a Hilbert space with inner product given by

$$\langle S,T\rangle = \sum_{i_1,\ldots,i_n=1}^{\infty} S(e_{i_1},\ldots,e_{i_n})\, T(e_{i_1},\ldots e_{i_n}).$$

In fact, it is easy to see by induction that $\mathcal{L}^n_{(2)}(H)$ is unitary equivalent to the Hilbert space of Hilbert-Schmidt operators from H into $\mathcal{L}^{n-1}_{(2)}(H)$.

Let (i,H,B) be an abstract Wiener space. Using Lemma 4.8 and Corollary 4.4, we see by induction that $\mathcal{L}^n(\underbrace{B,\ldots,B}_{n-1},H;\mathbb{R}) \subset \mathcal{L}^n_{(2)}(H)$. However, $\mathcal{L}^n(\underbrace{B,\ldots,B}_{n-2},H,H;\mathbb{R}) \not\subset \mathcal{L}^n_{(2)}(H)$, e.g., $S(x_1,\ldots x_{n-2},h_1,h_2) = (x_1,e)\ldots(x_{n-2},e)(h_1,h_2)$, $e \in B^*$, is in $\mathcal{L}^n(\underbrace{B,\ldots,B}_{n-2},H,H;\mathbb{R})$, but not in $\mathcal{L}^n_{(2)}(H)$. Moreover, Corollary 4.4 can be generalized to n-linear maps, i.e.

$$||T||_2 \leq a^{n-1} ||T||, \quad T \in \mathcal{L}^n(\underbrace{B,\ldots,B}_{n-1},H;\mathbb{R}),$$

where $a = \{\int_B ||x||^2 \, P_1(dx)\}^{1/2}$ and $||T|| = \sup\{|T(x_1,\ldots,x_{n-1},h)|;$

$x_1,\ldots, x_{n-1} \in B,\ h \in H,\ ||x_1|| = \cdots = ||x_{n-1}|| = |h| = 1\}$.

We do not have analogue of trace class operators for n-linear maps. But we have the following analogue for trace. Let H and K be two Hilbert spaces. Let S be a continuous bilinear map from $H \times H$ into K. S is said to be of <u>trace class type</u> if (i) $S_x \in \mathcal{L}_{(1)}(H)$ for all $x \in K$, where $\langle S_x h,k\rangle_H = \langle S(h,k),x\rangle_K$ and (ii) $x \mapsto \text{trace}_H S_x$ is a continuous linear functional in K. The definition implies that there is a unique vector, denoted by TRACE S, in K such that

$$\langle \text{TRACE } S,\ x\rangle_K = \text{trace}_H S_x,\ x \in K.$$

It is easy to see that TRACE S $= \sum_{k=1}^{\infty} S(e_k,e_k)$, where $\{e_k\}$ is an orthonormal basis of H. The following two results are obvious

(a) Let S be a continuous bilinear map from $H \times H$ into K such that for each orthonormal basis $\{e_k\}$ of H, the series $\sum_k |S(e_k,e_k)|_K$ is convergent and the supremum of $\sum_k |S(e_k,e_k)|_K$ over all orthonormal bases $\{e_k\}$ of H is finite. Then S is of trace class type.

(b) Let $T_1,T_2 \in \mathcal{L}_{(2)}(H)$ and S be a continuous bilinear map from $H \times H$ into K. Then $S\circ[T_1 \times T_2]$ is of trace class type.

<u>§2</u>. Let μ be a Gaussian measure in H with mean 0 and covariance operator S_μ. Let $\{e_k\}$ be an orthonormal basis of H given by the eigenvectors of S_μ with corresponding eigenvalues $\{\beta_k\}$. Observe that $\{\langle\cdot,e_k\rangle\}$ are independent Gaussian random variables in H

and $\langle \cdot , e_k \rangle$ is normally distributed with mean 0 and variance $\langle S_\mu e_k, e_k \rangle = \beta_k$. Let $\alpha > 0$ be such that $1-2\alpha\, \beta_k > 0$ for all k. Then

$$\int_H e^{\alpha |x|^2} \mu(dx) = \int_H e^{\sum_k \alpha \langle x, e_k \rangle^2} \mu(dx)$$

$$= \prod_k \int_H e^{\alpha \langle x, e_k \rangle^2} \mu(dx)$$

$$= \prod_k (1-2\alpha\, \beta_k)^{-1/2}$$

$$= [\prod_k (1-2\alpha\beta_k)]^{-1/2}.$$

Observe that the infinite product is convergent because $\sum_k \alpha\, \beta_k = \alpha \sum_k \beta_k < \infty$. Therefore, $\int_H e^{\alpha |x|^2} \mu(dx) < \infty$ when $0 < \alpha < \min_k \{1/2\beta_k\}$. However, for a Gaussian measure in a Banach space, this result is much harder to prove. See Fernique's theorem in Chapter III.

Theorem 2.4 can be generalized to any abstract Wiener space. Let U be a bounded linear operator from B into itself such that $U(H) \subset H$ and the restriction of U to H is a unitary operator of H. Then $p_t \circ U^{-1} = p_t$.

§3. The Wiener measure had been used by physicists long before it was shown to be actually a measure by N. Wiener. For the original proof of Theorem 3.1, see [49] or [50]. Our proof here is based on [8] and [51]. The proof of Theorem 3.4 is from [24]. For the detailed discussion of relation between function space integrals and partial differential equations, see [6].

Let w_c be the Wiener measure in $C[0,1]$ with parameter $c > 0$. It is easy to see that $w_c(E) = w(E/\sqrt{c})$, $E \in \mathcal{B}(C[0,1])$. Therefore, from Theorem 3.4 we have

$$\int_{C[0,1]} e^{-\alpha\int_0^1 x(t)^2 dt}\, w_c(dx) = \sqrt{\operatorname{sech}\sqrt{2\alpha c}}, \quad \alpha > 0.$$

This can be generalized to an abstract Wiener space (i,H,B). Let $S \in \mathcal{L}(B,H)$ such that $I+2t\alpha S(S|_H)^*$ is invertible, where $\alpha > 0$. Let

$$u(h) = \int_B \exp\{i\langle h, Sx\rangle - \alpha\,|Sx|^2\}\, p_t(dx), \quad h \in H.$$

Note that when $B = C[0,1]$, $h=0$, $t=c$ and $Sx(t)=\int_0^t x(s)ds$, we have the above case. In [34] it is shown that

$$u(0) = \{\det\,[I+2t\,\alpha\,S(S|_H)^*]\}^{1/2}$$

and

$$u(h) = u(0)\exp\{-t\,\langle [I+2t\alpha S(S|_H)^*]^{-1} S(S|_H)^* h, h\rangle/2\}.$$

§4. Let $H=L^2[0,1]$. A construction of the map n in Corollary 4.1 is given by the stochastic integral defined in §5, $n(f)(x) = \int_0^1 f(t)dx(t)$, where $f \in L^2[0,1]$ and $x \in C[0,1]$. For a general space, n is given by Lemma 4.7 and Theorem 6.1. In any case, n can be constructed by using Kolmogorov's theorem as we did in Proposition 4.2 and Corollary 4.1.

While this section is based on [18], it should be noted that we have tried very hard to provide simpler proofs than the origin[al] ones. The original proofs in [18] quote several theorems in [16]. We quote only one theorem (namely, Lemma 4.3) from [16]. A simpl[e] proof can be found in [9, p. 406].

The proof of Lemma 4.4(b) is very different from the original one. We use only finite dimensional version of Lemma 4.3. Corollary 4.2 is used in the proof of Theorem 6.3 and also in Chapter III. If B is a Hilbert space, then Corollary 4.2 is evident by Theorem 4.3. Recall that any $A \in \mathcal{L}_{(2)}(H)$ can be decomposed as $A = CK$, where $C \in \mathcal{L}_{(2)}(H)$ and $K \in \mathcal{K}(H)$. Therefore, Lemma 4.6 is obvious when $||\cdot||$ is Hilbertian.

One may conjecture that the operator K in Corollary 4.3 can be taken to be a Hilbert-Schmidt operator. The answer is negative. If this were the case, it would imply that for any measurable norm $||\cdot||$ in H and any orthonormal basis $\{e_n\}$ of H, there holds $\sum_n ||e_n||^2 < \infty$. This is false for the classical Wiener space. It is easy to see that $e_n(t) = \sqrt{2}\ \{1-[(n-\frac{1}{2})\pi]^{-1} \cos[(n-\frac{1}{2})\pi t]\}$, $n=1,2,\ldots$, is an orthonormal basis for C'. But $||e_n|| = \sqrt{2}$ for all n. Goodman also provides a counterexample. $||(x_1,\ldots,x_n,\ldots)|| = \sup_n n^{-\frac{1}{2}}|x_n|$ is a measurable norm in ℓ_2. Let $e_n = (0,\ldots,0,1,0\ldots)$, where 1 appears in the n-th place. Then $||e_n|| = n^{-\frac{1}{2}}$ and $\sum_i ||e_n||^2 = \infty$. However, the following is an open question: Let $||\cdot||$ be a measurable norm in H, does there exist an orthonormal basis $\{e_n\}$ of H such that $\sum_n ||e_n||^2 < \infty$? It is easy to see that this is true if $||\cdot||$ is Hilbertian.

The proof of Theorem 4.1 is same as the original one except that it has been notationally simplified here by embedding B* in B. In Chapter III we will present a probabilistic proof of this theorem due to Kallianpur. The proofs of Theorem 4.4 (in [18]) and Theorem 4.5 (in [19]) are same as the original ones.

We learned Theorem 4.6 through private conversations.

Corollary 4.4 has appeared in [34].

§5. Let Ω denote the probability space $C[0,1]$ with Wiener measure w. The elements of Ω will be denoted by ω. The stochastic proces given by $W(t,\omega) = \omega(t)$ is called a (one dimensional) Wiener process or Brownian motion. Ito's integral is an integral of the form

$$\int_0^1 f(t,\omega)dW(t,\omega),$$

where $f(t,\omega)$ is non-anticipating with respect to $W(t,\omega)$ and $\int_0^1 f(t,\omega)^2 dt < \infty$ almost surely. When f does not depend on ω, then Ito's integral reduces to the stochastic integral we defined in this section. Generalizations of Ito's integral to infinite dimensional Wiener process will be given in Chapter III.

The Wiener measure w in $C[0,1]$ extends to a Borel measure $\tilde{w}$ in $L^2[0,1]$ by

$$\tilde{w}(E) = w(E \cap C[0,1]), \quad E \in \mathcal{B}(L^2[0,1]).$$

It is easy to see that $\tilde{w}$ is a Gaussian measure in $L^2[0,1]$ with mean 0 and covariance operator S given by

$$S\, f(t) = \int_0^1 \min(t,s)\, f(s)\, ds.$$

Observe that $S = D\hat{D}$, where $D\, f(t) = \int_0^t f(s)ds$ and $\hat{D}$ is the adjoint of D in $L^2[0,1]$, i.e., $\hat{D}\, f(t) = \int_t^1 f(s)ds$. We saw before that $C' = D(L^2[0,1])$. It can be shown that $C' = \sqrt{S}(L^2[0,1])$, where $\sqrt{S}$ is the positive square root of S, i.e.,

$$\sqrt{S}\, f(t) = \sum_{n=1}^{\infty} 2[(n-\tfrac{1}{2})\pi]^{-1}[\int_0^1 f(s)\sin(n-\tfrac{1}{2})\pi s\, ds]\sin(n-\tfrac{1}{2})\pi t.$$

This idea of extending w to $\tilde{w}$ in order to capture C' is used by Kuelbs to study Gaussian measures in a general Banach space.

§6. If f satisfies the condition in Theorem 6.2 then it is uniformly continuous near zero in $\mathcal{T}_m$-topology (u.c.n.o. in $\mathcal{T}_m$). A function f in H is said to be u.c.n.o. in $\mathcal{T}_m$ if there exists a sequence $||\cdot||_n$ of measurable semi-norms such that $||\cdot||_n^{\sim}$ converges to zero in probability and f is uniformly continuous in $\mathcal{T}_m$-topology on $\{x \in H;\ ||x||_n \leq 1\}$ for each n. It is shown in [16] that if f is u.c.n.o. in $\mathcal{T}_m$ then the conclusion of Theorem 6.2 holds. Note that if we define $||x||_0 = \sum_{n=1}^{\infty} 2^{-n} E[||\cdot||_n^{\sim}]^{-1} ||x||_n$ then f is uniformly continuous in $\mathcal{T}_m$-topology on $\{x \in H;\ ||x||_0 \leq r\}$ for each $r > 0$. Thus f is u.c.n.o. in $\mathcal{T}_m$ if and only if there exists a measurable semi-norm $||\cdot||_0$ such that f is uniformly continuous in $\mathcal{T}_m$-topology on $\{x \in H;\ ||x||_0 \leq r\}$ for each $r > 0$.

Theorem 6.3 is taken from [18]. But the proof here is much simpler. Theorem 6.5 and Theorem 6.6 are infinite dimensional generalization of Lévy's continuity theorem. The proof of Theorem 6.5 is same as the original one. The original statement of Theorem 6.6 has $\mathcal{T}$-topology in stead of $\mathcal{T}_m$-topology. ($\mathcal{T}$-topology is defined to be the weakest topology on H for which $|A\,x|$ is continuous for all Hilbert-Schmidt operators A). Moreover, its proof is quite technical and complicated. Feldman gave a short proof of this theorem in [11]. We have not only put a weaker condition on ϕ (namely, we use $\mathcal{T}_m$-topology), but also simplified the proof. Theorem 6.7 was obtained independently by Gross [17] and Sazonov [40]. The original statement of this theorem uses also $\mathcal{T}$-topology. The proof here is taken from [17].

Chapter II. Equivalence and orthogonality of Gaussian measures.

In 1944 Cameron and Martin [3] discovered that w is quasi-invariant under translation by any function x, which is absolutely continuous and $x'(t)$ is of bounded variation. In 1950 Maruyama [36] found that by using stochastic integral (see Chapter I §5), it is sufficient to require that $x'(t)$ is in $L^2[0,1]$ (i.e. $x \in C'$, see Chapter I §5). In 1951 Sunouchi [45] and Cameron and Graves [2] gave independent proofs of Maruyama's theorem. Later on, we will give a formal proof of this theorem based on Donsker's "flat integral" [7]. Moreover, we will see that if w is translate by a function in $C \setminus C'$ then we obtain an orthogonal measure to w. This phenomenon is the so-called dichotomy of Gaussian measures. In fact we will prove that in an abstract Wiener space the translation measure of p_t by x is either equivalent or orthogonal to p_s and the equivalence occurs when and only when $x \in H$ and s=t.

In 1948, Kakutani obtained a very nice theorem for infinite product measures. More precisely, let $\mu_n \sim \nu_n$ (equivalent), n=1,2,3,.... then $\mu = \bigotimes_n \mu_n$ is either equivalent to $\nu = \bigotimes_n \nu_n$ or orthogonal. In 1958, Feldman and Hajek found independently th two Gaussian measures are either equivalent or orthogonal. In the same time, Segal proved this dichotomy theorem in a general situation for weak distributions. We will prove Feldman-Hajek's theorem for Gaussian measures in a Hilbert space (see Chapter I § based on Varadhan's lecture notes [46], and then we will construc Gaussian measures in a function space and give a simple proof of Feldman-Hajek's theorem due to Shepp. Finally, we will state som transformation formulas for an abstract Wiener space.

§1. Translation of Wiener measure.

Recall that if I is a cylinder set of the form

$$\{x\in C[0,1];\ (x(t_1),\ldots x(t_n))\in E\},\ 0 < t_1 < \ldots < t_n \leq 1,$$

then the Wiener measure W(I) of I is given by, ($t_o = u_o = 0$),

$$\left[\prod_{k=1}^{n} 2\pi\ (t_k - t_{k-1})\right]^{-1/2} \int_E \exp\left\{-\frac{1}{2}\sum_{k=1}^{n}\frac{(u_k-u_{k-1})^2}{t_k-t_{k-1}}\right\} du_1\ldots du_n.$$

Let us make an observation about the kernel function in the above integral: Write $u_k = x(t_k)$, k=1,2,...n, then

$$\frac{u_1^2}{t_1} + \frac{(u_2-u_1)^2}{t_2-t_1} + \ldots\ldots + \frac{(u_n-u_{n-1})^2}{t_n-t_{n-1}}$$

$$= \left[\frac{x(t_1)}{t_1}\right]^2 t_1 + \left[\frac{x(t_2)-x(t_1)}{t_2-t_1}\right]^2 (t_2-t_1) + \cdots + \left[\frac{x(t_n)-x(t_{n-1})}{t_n-t_{n-1}}\right]^2 (t_n-t_{n-1})$$

$$\approx x'(\tilde{t}_1)^2 t_1 + x'(\tilde{t}_2)^2 (t_2-t_1) + \ldots + x'(\tilde{t}_n)^2 (t_n-t_{n-1})$$

$$\approx \int_0^1 x'(t)^2\ dt.$$

Of course, the third and the fourth lines are just formal expressions since we know from Chapter I §3 that Brownian paths are nowhere differentiable. Nevertheless, we can regard the kernel formally as:

$$e^{-\frac{1}{2}\int_0^1 x'(t)^2\ dt}$$

and put

$$\delta x = e^{-\frac{1}{2}\int_0^1 x'(t)^2 dt}\ dx.$$

If f: C[0,1] → ℝ is w-integrable, we write

$$\int_{C[0,1]} f(x)\, w(dx) \text{ as } \rfloor\!\!\overline{} f(x)\ \delta x$$

<u>Definition 1.1</u>. $\rfloor\!\!\overline{} f(x)\ \delta x$ is called <u>Donsker's flat integral</u>.

Let $x_o \in C[0,1]$. Define the translation measure w_{x_o} of w by x_o by

$$w_{x_o}(E) = w(E + x_o), \quad E \in \mathcal{B}(C[0,1]).$$

<u>Theorem 1.1</u>. If $x_o \in C'$ then w_{x_o} is equivalent to w and the Radon-Nikodym derivative is given by

$$\frac{dw_{x_o}}{dw}(x) = e^{-\frac{1}{2}\int_0^1 x_o'(t)^2\, dt - \int_0^1 x_o'(t)\, dx(t)}$$

Remark. $\int_0^1 x_o'(t)dx(t)$ is regarded as a stochastic integral (Chapter I, §5) and dw_{x_o}/dw is defined w-almost everywhere. When Cameron-Martin first proved this formula, they regarded $\int_0^1 x_o'(t)dx(t)$ as a Stieltjes integral so that they assumed that x_o' is of bounded variation and thus dw_{x_o}/dw is defined everywhere in $C[0,1]$.

Proof. What we have to prove is that for any w-integrable function f, we have

$$\int_{C[0,1]} f(y)w(dy) = \int_{C[0,1]} f(x+x_o)g(x_o,x)w(dx),$$

where $g(x_o,x) = e^{-\frac{1}{2}\int_0^1 x_o'(t)^2 dt - \int_0^1 x_o'(t)\, dx(t)}$.

But, using Donsker's flat integral, we have

$$\int_{C[0,1]} f(y)w(dy) = \rfloor\!\!\overline{} f(y)\ \delta y$$

$$= \int\!\!\!\!\!\!\!\!\;\;\!\!\!\frown\; f(y) e^{-\frac{1}{2}\int_0^1 y'(t)^2\, dt} dy$$

$$= \int\!\!\!\!\!\!\!\!\;\;\!\!\!\frown\; f(x + x_o)\, e^{-\frac{1}{2}\int_0^1 (x'(t)+x_o'(t))^2\, dt} dx$$

$$= \int\!\!\!\!\!\!\!\!\;\;\!\!\!\frown\; f(x+x_o) e^{-\frac{1}{2}\int_0^1 x_o'(t)^2 dt - \int_0^1 x'_o(t)x'(t)dt}\, e^{-\frac{1}{2}\int_0^1 x'(t)^2\, dt} dx$$

$$= \int\!\!\!\!\!\!\!\!\;\;\!\!\!\frown\; f(x+x_o)\, e^{-\frac{1}{2}\int_0^1 x_o'(t)^2\, dt - \int_0^1 x_o'(t)dx(t)}\, \delta x$$

$$= \int_{C[0,1]} f(x+x_o)\, e^{-\frac{1}{2}\int_0^1 x_o'(t)^2\, dt - \int_0^1 x_o'(t)dx(t)}\, w(dx) \quad \#$$

In the case of an abstract Wiener space (i,H,B) we have the following theorem. Its proof can be regarded as an justification of the use of flat integral in the previous proof. For $x \in B$, define $p_t(x,E) = p_t(E+x)$, $E \in \mathcal{B}(B)$.

Theorem 1.2. If $h \in H$ then $p_t(h,\cdot)$ is equivalent to p_t and the Randon-Nikodym derivative is given by

$$\frac{dp_t(h,\cdot)}{dp_t}(x) = e^{-\frac{1}{2t}|h|^2 - \frac{1}{t}\langle h,x\rangle}, \quad x \in B.$$

Remark. $\langle h,x\rangle$ is regarded as a random variable over B. See Lemma 4.7.

Proof. We have to prove that for any p_t-integrable function f we have

$$\int_B f(y)\, p_t(dy) = \int_B f(x+h)\, e^{-\frac{1}{2t}|h|^2 - \frac{1}{t}\langle h,x\rangle}\, p_t(dx).$$

Obviously, it is sufficient to show the above equality for bounded continuous functions.

Let f be a bounded continuous function in B and $g = f|_H$. By Chapter I Theorem 6.3 $\tilde{g}$ makes sense and $\tilde{g} = f$ a.e. (p_t). Therefore we can pick up a sequence $\{P_n\} \subset \mathcal{F}$ converging to identity strongly such that $g(P_n x)^\sim$ converges to f a.e. (p_t). Hence

$$\int_B f(y)\, p_t(dy) = \lim_{n\to\infty} \int_B g(P_n\, x)^\sim\, (y)\, p_t\, (dy)$$

$$= \lim_{n\to\infty} \int_{P_n H} g(P_n\, x) \left(\frac{1}{\sqrt{2\pi t}}\right)^{\dim P_n H} e^{-\frac{1}{2t}|x|^2}\, dx$$

$$= \lim_{n\to\infty} \int_{P_n H} g(P_n x + P_n h) \left(\frac{1}{\sqrt{2\pi t}}\right)^{\dim P_n H} e^{-\frac{1}{2t}|x+h|^2}\, dx$$

$$= \lim_{n\to\infty} \int_{P_n H} \phi(P_n\, x) \left(\frac{1}{\sqrt{2\pi t}}\right)^{\dim P_n H} e^{-\frac{1}{2t}|x|^2}\, dx,$$

$$= \lim_{n\to\infty} \int_B \phi(P_n\, x)^\sim\, (y)\, p_t\, (dy),$$

where

$$\phi(y) = e^{-\frac{1}{2t}|h|^2 - \frac{1}{t}\langle y,h\rangle}\, g(y+h), \quad y \in H.$$

Clearly, $\phi(P_n\, x)^\sim \to e^{-\frac{1}{2t}|h|^2 - \frac{1}{t}\langle\cdot,h\rangle} f(\cdot+h)$ in probability.

Pick up a subsequence of P_n, if necessary, so that the above convergence is a.e. (p_t). Hence

$$\int_B f(y)p_t(dy) = \int_B f(y+h)e^{-\frac{1}{2t}|h|^2 - \frac{1}{t}\langle y,h\rangle} p_t(dy) \qquad \#$$

Corollary 1.1. (a) For each $\varepsilon > 0$, $w\{x \in C[0,1];\ ||x|| \le \varepsilon\} > 0$

(b) For each $\varepsilon > 0$, $p_t\{x \in B;\ ||x|| \le \varepsilon\} > 0$

Remark. This corollary implies that non-empty open sets have positive Wiener measure.

Proof. (a) See Chapter I Lemma 5.2 and Theorem 1.1 above.

(b) Use the same argument as in the proof of Chapter I Lemma 5.2 except that Theorem 1.2 should be used instead of Theorem 1.1. #

Corollary 1.2. If $\theta \in L^2[0,1]$ then $e^{-\int_0^1 \theta(t)dx(t)}$ is w-integrable and $\int_{C[0,1]} e^{-\int_0^1 \theta(t)dx(t)} w(dx) = e^{\frac{1}{2}\int_0^1 \theta(t)^2 dt}$.

Proof. Let $x_o(t) = \int_0^t \theta(s)\,ds$, then $x_o \in C'$. Put $f \equiv 1$ in Theorem 1.1 and obtain

$$1 = \int_{C[0,1]} e^{-\frac{1}{2}\int_0^1 \theta(t)^2 dt - \int_0^1 \theta(t)dx(t)} w(dx)$$

$$= e^{-\frac{1}{2}\int_0^1 \theta(t)^2 dt} \int_{C[0,1]} e^{-\int_0^1 \theta(t)dx(t)} w(dx) \qquad \#$$

§2. Kakutani's theorem on infinite product measures.

Let μ and ν be two probability measures in a measurable space $(X,\mathcal{B})$. Let λ be a probability measure such that both μ and ν are absolutely continuous with respect to λ. For instance, λ may be taken to be $\frac{1}{2}(\mu+\nu)$. Define the Hellinger integral of μ and as follows:

$$H(\mu,\nu) = \int_X \sqrt{\frac{d\mu}{d\lambda}} \sqrt{\frac{d\nu}{d\lambda}}\, d\lambda.$$

It can be checked easily that $H(\mu,\nu)$ is independent of the choice of λ and therefore it is often denoted by $\int_X \sqrt{d\mu d\nu}$. We have the following immediate properties for $H(\mu,\nu)$:

(i) $0 \leq H(\mu,\nu) \leq 1$, (by Schwarz's inequality),

(ii) $H(\mu,\nu) = 1 \iff \mu=\nu$,

(iii) $H(\mu,\nu) = 0 \iff \mu \perp \nu$ (orthogonal),

(iv) $\mu \sim \nu$ (equivalent) $\Rightarrow H(\mu,\nu) > 0$.

Exercise 28. Construct a counterexample for the converse of (iv).

Let $\Omega = \mathbb{R} \times \mathbb{R} \times \mathbb{R} \times \ldots\ldots$, and $\tilde{\mathcal{B}} = \bigotimes \mathcal{B}(\mathbb{R})$. We have a measurable space $(\Omega,\tilde{\mathcal{B}})$. Let $\{\mu_n\}$ and $\{\nu_n\}$ be two sequences of probability measures in $\mathbb{R}$. Define $\mu = \mu_1 \times \mu_2 \times \ldots\ldots$ and $\nu = \nu_1 \times \nu_2 \times \ldots$. Then μ and ν are two probability measures in $(\Omega,\tilde{\mathcal{B}})$. Obviously, if $\mu_k \perp \nu_k$ for some k, then $\mu \perp \nu$. But, how about $\mu_k \sim \nu_k$ for all k? See [26] for the proof of the following theorem.

Theorem 2.1. (Kakutani) If $\mu_k \sim \nu_k$, $k = 1,2,3\ldots$ then μ and ν are either equivalent or orthogonal. Moreover, $\mu \sim \nu$ if and only if $\prod_{k=1}^{\infty} H(\mu_k,\nu_k) > 0$. In case $\mu \sim \nu$, $\frac{d\mu}{d\nu} = \prod_{k=1}^{\infty} \frac{d\mu_k}{d\nu_k}$.

Example 1. Let $d\mu_k = \frac{1}{\sqrt{2\pi t}} e^{-\frac{x^2}{2t}} dx$ and $d\nu_k = \frac{1}{\sqrt{2\pi t}} e^{-\frac{(x+a_k)^2}{2t}} dx$.

Then $\mu \sim \nu$ if and only if $\sum_{k=1}^{\infty} a_k^2 < \infty$. In this case,

$$\frac{d\mu}{d\nu}\{(x_1,x_2,\ldots,x_n,\ldots)\} = e^{-\frac{1}{2t}\sum_{k=1}^{\infty} a_k^2 - \frac{1}{t}\sum_{k=1}^{\infty} a_k x_k}.$$

Proof. Use the easily checked formula $H(\mu_k,\nu_k) = e^{-\frac{a_k^2}{8t}}$. #

Example 2. Let $d\mu_k = \frac{1}{\sqrt{2\pi t}} e^{-\frac{x^2}{2t}} dx$ and $d\nu_k = \frac{1}{\sqrt{2\pi t}} e^{-\frac{(x+a_k)^2}{2s}} dx$, $k=1,2,\ldots$. If $t \neq s$ then $\mu \perp \nu$ no matter what a_k's are.

Proof. Use the easily checked formula

$$H(\mu_k,\nu_k) = [2\sqrt{ts}/(t+s)]^{1/2} e^{-a_k^2/4(t+s)}. \quad \#$$

Example 3. Let $d\mu_k = \frac{1}{\sqrt{2\pi t_k}} e^{-\frac{x^2}{2t_k}} dx$

and $d\nu_k = \frac{1}{\sqrt{2\pi t_k}} e^{-\frac{(x+a_k)^2}{2t_k}} dx$, $k = 1,2,3,\ldots$.

Then $\mu \sim \nu$ if and only if $\sum_{k=1}^{\infty} \frac{a_k^2}{t_k} < \infty$. In this case

$$\frac{d\mu}{d\nu}\{(x_1,\ldots x_n,\ldots)\} = e^{-\frac{1}{2}\sum_{k=1}^{\infty}\frac{a_k^2}{t_k} - \sum_{k=1}^{\infty}\frac{a_k x_k}{t_k}}.$$

Example 4. Let $d\mu_k = \frac{1}{\sqrt{2\pi}} e^{-\frac{x^2}{2}} dx$ and $d\nu_k = \frac{1}{\sqrt{2\pi t_k}} e^{-\frac{x^2}{2t_k}} dx$, $k \geq 1$.

If $\sum_{k=1}^{\infty} (t_k-1)^2 < \infty$ then $\mu \sim \nu$.

§3. Feldman-Hajek's theorem on equivalence of Gaussian measures in Hilbert space.

Let H be a real separable Hilbert space. In Chapter I §2 we showed that a Gaussian measure μ in H is uniquely determined by a vector m_μ of H and an $\mathcal{S}$-operator S_μ of H. m_μ is called the mean of μ and S_μ the covariance operator of μ. Feldman-Hajek's theorem asserts that any two Gaussian measures in H are either equivalent or orthogonal. For convenience, we use the notation $\mu = [a,S]$ to mean that μ is a Gaussian measure in H with mean a and covariance operator S. If $S \in \mathcal{S}$, $\sqrt{S}$ denotes the positive square root of S. Note that $\sqrt{S} \in \mathcal{L}_{(2)}(H)$ if $S \in \mathcal{S}$.

Theorem 3.1. Let $\mu = [0,S]$ and $\nu = [a,S]$. Then $\mu \sim \nu$ or $\mu \perp \nu$ according to $a \in \sqrt{S}(H)$ or not. Moreover, if $\mu \sim \nu$ then the Radon-Nikodym derivative is given by

$$\frac{d\nu}{d\mu}(x) = e^{-\frac{1}{2}|(\sqrt{S})^{-1} a|^2 + \langle(\sqrt{S})^{-1} a, (\sqrt{S})^{-1} x\rangle}$$

Remarks. (a) Regarding H as an abstract Wiener space $(i, \sqrt{S}(H), H)$ we see that a part of the above conclusion is Theorem 1.2 with $t = 1$ and $h = -a$.

(b) Note that $\int_H e^{i\langle b,x\rangle} d\mu(x) = e^{-\frac{1}{2}\langle Sb,b\rangle}$. Hence $\langle(\sqrt{S})^{-1}a, (\sqrt{S})^{-1}x\rangle$ is a random variable in H, normally distributed (w.r.t. μ) with mean 0 and variance $|(\sqrt{S})^{-1}a|^2$.

Proof. Step 1: μ not orthogonal to $\nu \Rightarrow a \in \sqrt{S}(H)$.

For each $h \in H$, let μ_h and ν_h be the distributions of $\langle h,\cdot\rangle$ with respect to μ and ν, respectively. Since μ is not orthogonal to ν, μ_h is not orthogonal to ν_h for any $h \neq 0$. Observe that μ_h is $N(0,\langle Sh,h\rangle)$ distributed and ν_h is $N(\langle a,h\rangle, \langle Sh,h\rangle)$ distributed. It follows that

$$\sup_{|h|=1} \frac{|\langle a,h\rangle|}{\sqrt{\langle Sh,h\rangle}} < \infty,$$

i.e., $a \in \sqrt{S}(H)$.

<u>Step 2:</u> $a \in \sqrt{S}(H) \Rightarrow \mu \sim \nu$, and $\frac{d\nu}{d\mu}(x)$ is given as in the theorem.

Regard H as an abstract Wiener space (i, H_o, H), where $H_o = \sqrt{S}(H)$ and the inner product of H_o is given by

$$\langle x,y\rangle_{H_o} = \langle(\sqrt{S})^{-1} x, (\sqrt{S})^{-1} y\rangle, \quad x,y \in H_o.$$

It is easy to see that μ is the Wiener measure p_1 of H and ν is $p_1(a,\cdot)$. Hence Theorem 1.2 gives the desired conclusion. #

<u>Corollary 3.1</u> Let $\mu = [a,S]$ and $\nu = [b,S]$. Then $\mu \sim \nu$ or $\mu \perp \nu$ according to $a-b \in \sqrt{S}(H)$ or not.

Proof. Simply note that equivalence and orthogonality are preserved by translations. #

<u>Exercise 29.</u> Find the Radon-Nikodym derivative $d[a,S]/d[b,S]$.

<u>Theorem 3.2.</u> Let $\mu = [0,S_1]$ and $\nu = [0,S_2]$. If μ is not orthogonal to ν then $S_2 = \sqrt{S_1}\, T \sqrt{S_1}$, where T is a positive definite, bounded, invertible operator and $T - I \in \mathcal{L}_{(2)}(H)$.

Proof. Let $\lambda_1, \lambda_2, \ldots\ldots, \lambda_n, \ldots\ldots$ be non-zero eigenvalues of S_1 and let $e_1, e_2, \ldots\ldots, e_n \ldots\ldots$ be the corresponding unit eigenvectors. Define $\mathcal{H} = \{(s_1, s_2, \ldots\ldots);\ \sum_{j=1}^{\infty} \lambda_j\, s_j^2 < \infty\}$ and a map $\Phi\colon H \to \mathcal{H}$ by

$$\Phi(x) = (\langle x, e_1\rangle/\sqrt{\lambda_1}, \ldots\ldots, \langle x, e_n\rangle/\sqrt{\lambda_n}, \ldots\ldots).$$

Observe that Φ is onto and $\ker \Phi = \ker S_1$. Let $\tilde{\mu}$ and $\tilde{\nu}$ be the distributions of μ and ν on $\mathcal{H}$. $(\Phi(x))_i$ will denote the i-th component of $\Phi(x)$. Clearly, for $i,j = 1,2,\ldots$, we have

$$\int_H (\Phi(x))_i\, d\mu(x) = \int_H (\Phi(x))_i\, d\nu(x) = 0,$$

$$\int_H (\Phi(x))_i (\Phi(x))_j\, d\mu(x) = \delta_{ij},$$

and

$$\int_H (\Phi(x))_i (\Phi(x))_j\, d\nu(x) = \frac{\langle S_2 e_i, e_j\rangle}{\sqrt{\lambda_i \lambda_j}} \equiv t_{ij}$$

That is,

$$\int_{\mathcal{H}} s_i\, d\tilde{\mu}(s) = \int_{\mathcal{H}} s_i\, d\tilde{\nu}(s) = 0$$

$$\int_{\mathcal{H}} s_i s_j\, d\tilde{\mu}(s) = \delta_{ij}$$

$$\int_{\mathcal{H}} s_i s_j\, d\tilde{\nu}(s) = t_{ij}$$

Since μ and ν are not orthogonal, $\tilde{\mu}$ and $\tilde{\nu}$ are not orthogonal. Let $\tilde{\mu}_n$ and $\tilde{\nu}_n$ be the restrictions of $\tilde{\mu}$ and $\tilde{\nu}$ to the first n coordinates. Note tbat $\tilde{\mu}_n$ is Gaussian with mean 0 and covariance operator I and $\tilde{\nu}_n$ another Gaussian measure with mean 0 and covariance operator T_n, where

$$T_n = \begin{pmatrix} t_{11} & t_{12} \cdots t_{1n} \\ \cdots & \cdots \\ t_{n1} & t_{n2} \cdots t_{nn} \end{pmatrix}.$$

Exercise 30. Let λ_1 and λ_2 be two Gaussian measures in $\mathbb{R}_n$ with means a_1 and a_2, and covariance operators S_1 and S_2, respectively. Then the Hellinger integral of λ_1 and λ_2 is given by

$$H(\lambda_1,\lambda_2)^2 = \left\{\frac{\det S_1 \det S_2}{\det(S_1+S_2)/2}\right\}^{1/2} \exp\left\{-\langle (S_1+S_2)^{-1}(a_1-a_2), a_1-a_2\rangle\right\}.$$

Now, let $\Theta_{n,j}$, j=1,...,n, be the eigenvalues of T_n. Then by the above exercise we have

$$2 \log H(\tilde{\mu}_n,\tilde{\nu}_n) = \sum_{j=1}^{n}\left[\frac{1}{2}\log \Theta_{n,j} - \log\left(\frac{1+\Theta_{n,j}}{2}\right)\right].$$

Since $\tilde{\mu}$ is not orthogonal to $\tilde{\nu}$, we have $H(\tilde{\mu},\tilde{\nu}) > 0$. Observe that $H(\tilde{\mu}_n,\tilde{\nu}_n)$ monotonically decreases to $H(\tilde{\mu},\tilde{\nu})$ as $n \to \infty$. Therefore, there exists a finite constant M such that $H(\tilde{\mu}_n,\tilde{\nu}_n) > e^{-4M}$ for all n. That is

$$\sum_{j=1}^{n}\left[2\log\left(\frac{1+\Theta_{n,j}}{2}\right) - \log \Theta_{n,j}\right] < M \text{ for all } n.$$

This implies that

a) $\sum_{j=1}^{n}(1-\Theta_{n,j})^2 < M$ for all n, (since $(1-x)^2 \le 2\log\frac{1+x}{2} - \log x$)

b) there exists $0 < c < \infty$ such that $c < \Theta_{n,j} < \frac{1}{c}$ for all n,j.

Define an operator $T: H \to H$ by requiring $Tx = x$ when $x \in \ker S_1$ and $Te_i = \sum_{j=1}^{\infty} t_{ij}\, e_j$, i = 1,2,... T is clearly positive definite.

(b) implies that T is bounded and invertible.

Moreover,

$$\sum_{i,j=1}^{\infty} (t_{ij}-\delta_{ij})^2 = \lim_{n\to\infty} \sum_{i,j=1}^{n} (t_{ij} - \delta_{ij})^2$$

$$= \lim_{n\to\infty} \sum_{j=1}^{n} (1 - \Theta_{n,j})^2$$

$$< M \text{ (by (a))}.$$

Hence, $T-I \in \mathcal{L}_{(2)}(H)$.

<u>Exercise 31</u>. If $[0,S_1]$ is not orthogonal to $[0,S_2]$ then $\ker S_1 = \ker S_2$.

By the above exercise, we have $\sqrt{S_1}\, T \sqrt{S_1} = S_2$ in $\ker S_1$. Moreover, for $i = 1,2,\ldots,$

$$\sqrt{S_1}\, T \sqrt{S_1}\, e_i = \sqrt{S_1}\, T (\sqrt{\lambda_i}\, e_i)$$

$$= \sqrt{S_1} \left(\sum_{j=1}^{\infty} \sqrt{\lambda_i}\, t_{ij}\, e_j\right)$$

$$= \sum_{j=1}^{\infty} \sqrt{\lambda_i} \sqrt{\lambda_j}\, t_{ij}\, e_j$$

$$= \sum_{j=1}^{\infty} \langle S_2 e_i, e_j\rangle e_j$$

$$= S_2 e_i.$$

Hence we have $\sqrt{S_1}\, T \sqrt{S_1} = S_2$ in H. #

<u>Theorem 3.3</u> Let $\mu = [0,S_1]$ and $\nu = [0,S_2]$. Suppose $S_2 = \sqrt{S_1}\ T\ \sqrt{S_1}$, where T is positive definite, bounded, invertible and $T-I \in \mathcal{L}_{(2)}(H)$. Then μ is equivalent to ν. Moreover, the Radon-Nikodym derivative is given by:

$$\frac{d\nu}{d\mu}(\psi\ \{x_n\}) = \prod_{k=1}^{\infty} \frac{1}{\sqrt{\beta_k+1}}\, e^{\frac{\beta_k}{2(\beta_k+1)} x_k^2},$$

where ψ and β_k are given in the following proof.

Remark. Restrict our attention to $(\ker S_1)^{\perp}$, if necessary, we may assume that S_1 is one-to-one. We can then regard H as an abstract Wiener space $(i, \sqrt{S_1}(H), H)$ as before, then μ is the Wiener measure p_1 of H. Moreover, it is easy to see that $\nu = p_1 \circ L^{-1}$, where $L = \sqrt{S_2}(\sqrt{S_1})^{-1}$. If S_1 and S_2 are related as in the theorem then L has the same properties as T. Therefore, the above theorem can be considered as a special case of transformation formula to be discussed in §6.

Proof. Let $\{e_k\}$ be an orthonormal basis of H consisting of eigenvectors of T-I and let $\{\beta_k\}$ be the corresponding eigenvalues. Then

$$Te_k = (\beta_k + 1)e_k, \quad k=1,2,\ldots.$$

Let $\Omega = \mathbb{R} \times \mathbb{R} \times \ldots\ldots$ and $\tilde{\mu} = \mu_1 \times \mu_2 \times \ldots$, $\tilde{\nu} = \nu_1 \times \nu_2 \times \ldots$,

where

$$\mu_k = \frac{1}{\sqrt{2\pi}}\, e^{-\frac{x^2}{2}}\, dx \quad \text{for all } k$$

and $\nu_k = \frac{1}{\sqrt{2\pi(\beta_k+1)}} e^{-\frac{x^2}{2(\beta_k+1)}} dx \qquad k=1,2,\ldots .$

Since $\sum_{k=1}^{\infty} \beta_k^2 < \infty$, $\tilde{\mu}$ is equivalent to $\tilde{\nu}$ by §2 Example 4 and

$$\frac{d\tilde{\nu}}{d\tilde{\mu}}((x_1,x_2,\ldots.)) = \prod_{k=1}^{\infty} \frac{1}{\sqrt{\beta_k+1}} e^{\frac{\beta_k}{2(\beta_k+1)} x_k^2} .$$

Exercise 32. Show that there is a subspace Ω_o of Ω such that $\tilde{\mu}(\Omega_o) = \tilde{\nu}(\Omega_o) = 1$ and the above infinite product is convergent when $(x_1,x_2,\ldots) \in \Omega_o$. Show also that when $T-I \in \mathcal{L}_{(1)}(H)$, $\frac{d\tilde{\nu}}{d\tilde{\mu}}$ can be expressed as

$$\frac{d\tilde{\nu}}{d\tilde{\mu}}((x_1,x_2,\ldots)) = (\prod_{k=1}^{\infty}(\beta_k+1))^{-1/2} e^{\frac{1}{2}\sum_{k=1}^{\infty}\frac{\beta_k}{\beta_k+1} x_k^2}.$$

Define a map $\psi : \Omega \to H$ by

$$\psi(x) = \sum_{k=1}^{\infty} x_k \sqrt{S_1}\ e_k, \quad x = (x_1,x_2\ldots) \in \Omega .$$

It is easy to check that $\int_\Omega |\psi(x)|^2 \tilde{\mu}(dx) = \text{trace } S_1$ and $\int_\Omega |\psi(x)|^2 \tilde{\nu}(dx) = \text{trace } S_1 T = \text{trace } \sqrt{S_1}\ T \sqrt{S_1} = \text{trace } S_2$. Hence $\psi(x) \in H$ almost surely with respect to both $\tilde{\mu}$ and $\tilde{\nu}$. Such a map ψ is often called a random vector.

We show that $\mu = \tilde{\mu}\circ\psi^{-1}$ and $\nu = \tilde{\nu}\circ\psi^{-1}$ as follows. Note that $\tilde{\mu}\circ\psi^{-1}$ and $\tilde{\nu}\circ\psi^{-1}$ are Gaussian with mean 0. Moreover,

$$\int_H \langle x,z\rangle \langle y,z\rangle\ \tilde{\mu}\circ\psi^{-1}(dz)$$

$$= \int_H \langle x,\psi(z)\rangle \langle y,\psi(z)\rangle\ \tilde{\mu}\ (dz)$$

$$= \int_\Omega \sum_{j,k=1}^{\infty} z_j\, z_k \langle \sqrt{S_1}\, e_j, x\rangle \langle \sqrt{S_1}\, e_k, y\rangle \quad \tilde{\mu}(dz)$$

$$= \sum_{j,k=1}^{\infty} \int_\Omega z_j\, z_k \langle \sqrt{S_1}\, e_j, x\rangle \langle \sqrt{S_1}\, e_k, y\rangle \quad \tilde{\mu}(dz)$$

$$\left(\text{Since } \sum_{j=1}^{\infty} z_j^2 \langle \sqrt{S_1}\, e_j, x\rangle^2 \in L^1(\Omega, \tilde{\mu})\right)$$

$$= \langle \sqrt{S_1}\, x, \sqrt{S_1}\, y\rangle$$

$$= \langle S_1 x, y, \rangle.$$

Similarly,

$$\int_H \langle x,z\rangle \langle y,z\rangle\, \tilde{\nu} \circ \psi^{-1}(dz) = \langle S_2 x, y\rangle.$$

Therefore, we have $\mu = \tilde{\mu} \circ \psi^{-1}$ and $\nu = \tilde{\nu} \circ \psi^{-1}$. It follows that $\mu \sim \nu$. From the above formula for $\frac{d\tilde{\nu}}{d\tilde{\mu}}$ and the relation $\frac{d\nu}{d\mu}(\psi(x)) = \frac{d\tilde{\nu}}{d\tilde{\mu}}(x)$, we obtain easily our forumla in the theorem #

Lemma 3.1. If $[a_1, S_1]$ and $[a_2, S_2]$ are not orthogonal then $[0,S_1]$ and $[0,S_2]$ are not orthogonal.

Proof. An easy application of Exercise 30. #

Theorem 3.4. $[a_1,S_1]$ and $[a_2,S_2]$ are either equivalent or orthogonal. They are equivalent if and only if

(a) $[a_1,S_1]$ and $[a_2,S_1]$ are equivalent, and

(b) $[a_2,S_1]$ and $[a_2,S_2]$ are equivalent.

Remark. The necessary and sufficient condition for (a) is given in Corollary 3.1, while that for (b) is given in Theorem 3.3

Proof. Suppose $[a_1,S_1]$ is not orthogonal to $[a_2,S_2]$ then by Lemma 3.1 $[0,S_1]$ is not orthogonal to $[0,S_2]$. Therefore, by Theorem 3.2 and Theorem 3.3, $[0,S_1]$ is equivalent to $[0,S_2]$. Therefore, $[a_2,S_1]$ and $[a_2,S_2]$ are equivalent. Hence, $[a_1,S_1]$ is not orthogonal to $[a_2,S_1]$. By Corollary 3.1, $[a_1,S_1]$ and $[a_2,S_1]$ are equivalent. Therefore, $[a_1,S_1]$ and $[a_2,S_2]$ are equivalent. #

§4. Equivalence and orthogonality of Gaussian measures in function space.

Suppose $x_0 \in \mathbb{R}^n$ and $A = (a_{ij})$ is a strictly positive definite n-by-n matrix. $|A|$ will denote the determinant of A. Since A is assumed to be strictly positive definite, we have $|A| > 0$. Corresponding to x_0 and A, we define a Gaussian measure μ in $\mathbb{R}^n$ by:

$$\mu(dx) = \frac{1}{(\sqrt{2\pi})^n \sqrt{|A|}} e^{-\frac{1}{2}\langle A^{-1}(x-x_0),\, x-x_0\rangle} dx,$$

where A^{-1} is the inverse of A, and $\langle , \rangle$ is the Euclidean inner product, dx is the Lebesgue measure of $\mathbb{R}^n$. The following lemma is easy to check.

Lemma 4.1. (a) $\int_{\mathbb{R}^n} \langle y, x\rangle \mu(dx) = \langle y, x_0\rangle$,

(b) $\int_{\mathbb{R}^n} \langle y, x\rangle\langle z, x\rangle \mu(dx) = \langle Ay, z\rangle$,

(c) $\hat{\mu}(y) = e^{i\langle y, x_0\rangle - \frac{1}{2}\langle Ay, y\rangle}$.

In case A is assumed to be positive definite, then $|A|$ can be zero and thus we can not define μ by the above formula. In this case, we define μ in terms of its Fourier transform in Lemma 4.1 (c) and then apply Bochner's theorem. For instance, if A=0 then μ is the δ measure concentrating at x_0. In any case, corresponding to $x_0 \in \mathbb{R}^n$ and a positive definite n-by-n matrix $A = (a_{ij})$ we have a unique Gaussian measure μ satisfying Lemma 4.1.

Let Λ be any index set. Let $\Omega = \mathbb{R}^{\Lambda}$ be the space consisting of real-valued functions ω defined in Λ. Let $\mathcal{B}$ be the smallest σ-field in Ω such that each coordinate function $\omega(t)$, $t \in \Lambda$, is measurable.

Definition 4.1. $\rho : \Lambda \times \Lambda \to \mathbb{R}$ is called a covariance function if for any finite set $\pi = \{t_1, \cdots, t_k\} \subset \Lambda$, the matrix $\rho^{\pi} = (\rho(t_i, t_j))$ is positive definite.

Let $m \in \Omega$ and ρ be a covariance function. For each finite subset π of Λ, we can associate a Gaussian measure μ^{π} using m^{π} and ρ^{π}, where $m^{\pi}(t_1, \cdots, t_k) = (m(t_1), \cdots, m(t_k))$ and $\rho^{\pi} = (\rho(t_i, t_j))$. It is easy to check that $\{\mu^{\pi};\ \pi \subset \Lambda \text{ finite}\}$ is a consistent family of measures. Therefore, by Kolmogorov's extension theorem, there is a unique measure μ in Ω such that

$$\mu\{\omega\ ;\ (\omega(t_1), \cdots, \omega(t_k)) \in E\} = \mu^{\pi}(E),\ \pi = \{t_1, \cdots, t_k\}.$$

This measure μ is called the Gaussian measure corresponding to the mean funciton m and the covariance funciton ρ.

Example 1. $\Lambda = [0, 1]$. Take $m=0$ and $\rho(t, s) = \min(t, s)$. The corresponding Gaussian measure concentrates in $C[0, 1]$ and is the Wiener measure discussed in Chapter I §3.

Example 2. $\Lambda = [0, 1]$. Take $m=0$ and ρ defined by $\rho(t, s)=1$ if $t=s$ and $\rho(t, s)=0$ if $t \neq s$. The corresponding Gaussian measure is a σ-additive extension of the Gauss measure in $L^2[0, 1]$ and concentrates in any abstract Wiener space $(i, L^2[0, 1], B)$, where $B \subset \mathbb{R}^{[0,1]}$.

Let us recall that if μ and υ are two probability measures in a measurable space $(\Omega, \mathcal{B})$, then we can define the Hellinger integral of μ and υ by:

$$H(\mu, \upsilon) = \int_\Omega \sqrt{d\mu d\upsilon}.$$

We define another functional, the so-called J-functional, as follows: Let $\xi = \mu + \upsilon$ and define

$$J(\mu, \upsilon) = \int_\Omega \left(\frac{d\mu}{d\xi} - \frac{d\upsilon}{d\xi}\right) \log\left(\frac{d\mu}{d\xi} \Big/ \frac{d\upsilon}{d\xi}\right) d\xi.$$

By convention the integrand of J is taken to be ∞ if either $\frac{d\mu}{d\xi}$ is 0 or $\frac{d\upsilon}{d\xi}$ is 0, but not both. $(\frac{d\mu}{d\xi} + \frac{d\upsilon}{d\xi} = 1)$.

Properties for J:

(a) $0 \leq J(\mu, \upsilon) \leq \infty$, and

(b) $J(\mu, \upsilon) < \infty \Rightarrow \mu \sim \upsilon$.

Relation between H and J:

(a) $J < \infty \Rightarrow H > 0$ $(J < \infty \Rightarrow \mu \sim \upsilon \Rightarrow H > 0)$,

(b) $H > 0$ does not imply that $J < \infty$.

Exercise 33. Construct a conterexample for (b).

Our object in this section is to show that if μ and υ are two Gaussian measures in a function space $(\Omega, \mathcal{B})$ then we do have $H > 0 \Longrightarrow J < \infty$. In order to prove this assertion, we need Helms' martingale convergence theorem [22].

Let $(\mathcal{D}, <)$ be a directed set. Let $(\Omega, \mathcal{B}, \mu)$ be a finite measure space. Suppose for each $a \in \mathcal{D}$, we have a sub-σ-field $\mathcal{B}_a$ of $\mathcal{B}$, such that $a < b \Rightarrow \mathcal{B}_a \subset \mathcal{B}_b$. We will assume, without loss of generality, that $\sigma[\bigcup_{a \in \mathcal{D}} \mathcal{B}_a] = \mathcal{B}$.

<u>Definition 4.2</u>. A net $\{x_a; a \in \mathcal{D}\}$ in $L^1(\Omega)$ is called a <u>martingale</u> with respect to $\{\mathcal{B}_a; a \in \mathcal{D}\}$ if

(a) x_a is measurable with respect to $\mathcal{B}_a$ for all $a \in \mathcal{D}$, and

(b) $E[x_b | \mathcal{B}_a] = x_a$, whenever $a < b$.

<u>Theorem 4.1 (Helms)</u>. Let $\{x_a; a \in \mathcal{D}\}$ be a martingale in $L^1(\Omega)$. Then the following statements are equivalent:

(a) The net $\{x_a; a \in \mathcal{D}\}$ is uniformly integrable.

(b) There exists $x_\infty \in L^1(\Omega)$ such that $x_a = E[x_\infty | \mathcal{B}_a]$ for all $a \in \mathcal{D}$

(c) The net $\{x_a; a \in \mathcal{D}\}$ converges strongly in $L^1(\Omega)$.

Remark. If for some $p > 1$, $x_a \in L^p(\Omega)$ for all a, then (b) $\Leftrightarrow$ (c) can be strengthened to $L^p(\Omega)$ in stead of $L^1(\Omega)$.

Proof. Since we will use only the implication (b) $\Rightarrow$ (c), we wil prove this assertion only. Since $\mathcal{B} = \sigma[\bigcup_{a \in \mathcal{D}} \mathcal{B}_a]$, given any $\varepsilon > 0$, there exists $x_\varepsilon = \sum_{i=1}^{n} \tau_i 1_{E_i}$, where $\tau_i \in \mathbb{R}$ and $E_i \in \bigcup_{a \in \mathcal{D}} \mathcal{B}_a$, such that $\|x_\infty - x_\varepsilon\|_1 < \varepsilon/_2$. Clearly,

$$\|x_\infty - x_a\|_1 = \|x_\infty - E[x_\infty|\mathcal{B}_a]\|_1$$

$$= \|(x_\infty - x_\varepsilon)+x_\varepsilon - E[(x_\infty - x_\varepsilon)+ x_\varepsilon|\mathcal{B}_a]\|_1$$

$$\leq \|x_\infty - x_\varepsilon\|_1 + \|x_\varepsilon - E[x_\varepsilon|\mathcal{B}_a]\|_1 + \|E[x_\infty - x_\varepsilon|\mathcal{B}_a]\|_1 .$$

But

$$\|E[x_\infty - x_\varepsilon|\mathcal{B}_a]\|_1 = \int_\Omega |E[x_\infty - x_\varepsilon|\mathcal{B}_a]|d\mu$$

$$\leq \int_\Omega E[|x_\infty - x_\varepsilon||\mathcal{B}_a]d\mu$$

$$= \int_\Omega |x_\infty - x_\varepsilon|d\mu$$

$$= \|x_\infty - x_\varepsilon\|_1 .$$

Moreover, there exists $a_\varepsilon \in \mathcal{D}$ such that $E_i \in \mathcal{B}_{a_\varepsilon}$, i=1,2,···, n. Hence if $a > a_\varepsilon$ then $E[x_\varepsilon|\mathcal{B}_a]=x_\varepsilon$. Therefore, whenever $a > a_\varepsilon$, we have

$$\|x_\infty - x_a\|_1 \leq 2\|x_\infty - x_\varepsilon\|_1 < \varepsilon. \qquad \#$$

Now, let us fix two Gaussian measures μ_1 and μ_2 in a function space $\Omega = \mathbb{R}^\Lambda$. Let m_i and ρ_i be the mean function and covariance function of μ_i, i=1,2. Let $\mu = \mu_1 + \mu_2$. Let $\mathcal{D}$ consist of all finite subsets of Λ. $\mathcal{D}$ is a directed set with partial order given by set inclusion. Recall that if $\pi = \{t_1, \cdots, t_k\} \in \mathcal{D}$ then

$$m_i^\pi (t_1, \cdots, t_k) = (m_i(t_1), \cdots, m_i(t_k))$$

and $\rho_i^\pi = (\rho_i(t_j, t_k))$.

For each $\pi \in \mathcal{D}$, let $\mathcal{B}_\pi$ be the σ-field generated by the coordinate functions $\omega(t_1), \cdots, \omega(t_k)$. Let μ_i^π be the restriction of μ_i to $\mathcal{B}_\pi$.

Lemma 4.2. Let $x_i^\pi \equiv \frac{d\mu_i^\pi}{d\mu}$, $i=1,2$. Then $\{x_i^\pi ; \pi \in \mathcal{D}\}$ is a martingale net in $L^1(\Omega, \mu)$.

Proof. In fact $x_i^\pi = E\{ \frac{d\mu_i}{d\mu} | \mathcal{B}_\pi \}$. #

Lemma 4.3. $x_i^\pi \to x_i \equiv \frac{d\mu_i}{d\mu}$ in $L^1(\Omega, \mu)$.

Proof. By Theorem 4.1. #

Define $H(\pi) = \int_\Omega \sqrt{x_1^\pi \cdot x_2^\pi}\, d\mu$ and $J(\pi) = \int_\Omega (x_1^\pi - x_2^\pi) \log \frac{x_1^\pi}{x_2^\pi}\, d\mu$.

Lemma 4.4. (a) $H(\pi_1) \le H(\pi_2)$ if $\pi_1 > \pi_2$,

(b) $J(\pi_1) \ge J(\pi_2)$ if $\pi_1 > \pi_2$,

(c) $H(\mu_1, \mu_2) = \inf_{\pi \in \mathcal{D}} H(\pi)$,

(d) $J(\mu_1, \mu_2) = \sup_{\pi \in \mathcal{D}} J(\pi)$.

Proof. (a) $H(\pi_1) = E_\mu(\sqrt{x_1^{\pi_1}\, x_2^{\pi_1}})$

$$= E_\mu[E[\sqrt{x_1^{\pi_1} x_2^{\pi_1}} | \mathcal{B}_{\pi_2}]]$$

$$\leq E_\mu\{E[x_1^{\pi_1} | \mathcal{B}_{\pi_2}] E[x_2^{\pi_1} | \mathcal{B}_{\pi_2}]\}^{1/2}$$

$$= E_\mu(x_1^{\pi_2} x_2^{\pi_2})^{1/2} \qquad \text{(by Lemma 4.2)}$$

$$= H(\pi_2).$$

(b) Use the following exercise and note that $(x - y)(\log x - \log y)$ is a convex funciton in x and y.

Exercise 34. If $\Theta(x, y)$ is a convex function in $\mathbb{R}^2$ then

$$\Theta(E[f|\mathcal{B}], E[g|\mathcal{B}]) \leq E(\Theta(f, g)|\mathcal{B}).$$

(c) and (d) follow from (a), (b) and Lemma 4.3. #

Lemma 4.5. Suppose ρ_i, $i=1,2$, are strictly positive definite, i.e. ρ_i^π are strictly positive definite for all $\pi \in \mathcal{D}$. Then

$$H(\pi)^2 = (\det \rho_1^\pi \det \rho_2^\pi)^{\frac{1}{2}} \left(\det \frac{\rho_1^\pi + \rho_2^\pi}{2}\right)^{-1} \exp\{-\langle(\rho_1^\pi + \rho_2^\pi)^{-1}(m_1^\pi - m_2^\pi), m_1^\pi - m_2^\pi\rangle\}$$

and

$$J(\pi) = \frac{1}{2}\,\mathrm{trace}\{(\rho_2^\pi - \rho_1^\pi)((\rho_1^\pi)^{-1} - (\rho_2^\pi)^{-1})\}$$

$$+ \frac{1}{2}\langle((\rho_1^\pi)^{-1} + (\rho_2^\pi)^{-1})(m_1^\pi - m_2^\pi), m_1^\pi - m_2^\pi\rangle.$$

Proof. The first formula follows from Exercise 30, while the second formula can be checked easily by direct computation. #

Notation for discussion.

$$D(\pi)= (\det \frac{\rho_1^\pi+\rho_2^\pi}{2})^2 (\det \rho_1^\pi \det\rho_2^\pi)^{-1},$$

$$E(\pi)= <(\frac{\rho_1^\pi+\rho_2^\pi}{2})^{-1}(m_1^\pi - m_2^\pi),\ m_1^\pi - m_2^\pi>.$$

Then obviously, we have $H(\pi)^{-4}= D(\pi)e^{E(\pi)}$.

Lemma 4.6. If $H(\mu_1, \mu_2) > 0$ then there exists a constant M independent of π such that

$$D(\pi) \le M, \quad E(\pi) \le M \quad \text{for all} \quad \pi \in \mathcal{D}.$$

Proof. If we let $m_1 = m_2$ then $H(\pi)^{-4}= D(\pi) \ge 1$ since $H(\pi) \le 1$. Observe that $D(\pi)$ is independent of m_i, we must have $D(\pi) \ge 1$ in the general case. Also $e^{E(\pi)} \ge 1$ since $E(\pi) \ge 0$.

By the assumption that $H(\mu_1, \mu_2) > 0$ and by Lemma 4.4(c), there exists $\delta > 0$ such that $H(\pi) \ge \delta$ for all $\pi \in \mathcal{D}$. Hence

$$D(\pi)= H(\pi)^{-4} e^{-E(\pi)} \le \delta^{-4},$$

and
$$E(\pi)= -4 \log H(\pi) - \log D(\pi)$$
$$\le -4 \log \delta. \qquad \#$$

Notation for discussion.

$$T(\pi) = \text{trace}\{(\rho_2^\pi - \rho_1^\pi)((\rho_1^\pi)^{-1} - (\rho_2^\pi)^{-1})\},$$

$$Q(\pi) = \langle((\rho_1^\pi)^{-1} + (\rho_2^\pi)^{-1})(m_1^\pi - m_2^\pi),\ m_1^\pi - m_2^\pi\rangle.$$

Then we have $2J(\pi) = T(\pi) + Q(\pi)$.

Lemma 4.7. If $H(\mu_1, \mu_2) > 0$ then there exists a constant N independent of π such that

$$T(\pi) \le N \quad \text{and} \quad Q(\pi) \le N \quad \text{for all } \pi \in \mathcal{D}.$$

Proof. (a) Let $\lambda_i > 0$, $i=1,2,\cdots,n$ be the eigenvalues of $\rho_1^\pi(\rho_2^\pi)^{-1}$ (π has n points). Then

$$T(\pi) = \sum_{i=1}^{n} (\lambda_i + \frac{1}{\lambda_i} - 2) = \sum_{i=1}^{n} \frac{(\lambda_i - 1)^2}{\lambda_i},$$

and

$$D(\pi) = \prod_{i=1}^{n} \frac{1+\lambda_i}{2} \cdot \frac{1+\frac{1}{\lambda_i}}{2}.$$

Using the inequality $a_1 + a_2 + \cdots + a_n \le 4(1+\frac{a_1}{4})(1+\frac{a_2}{4})\cdots\cdots(1+\frac{a_n}{4})$, $a_i \ge 0$, we have for all $\pi \in \mathcal{D}$,

$$T(\pi) \le 4 \prod_{i=1}^{n} (1 + \frac{(\lambda_i - 1)^2}{4\lambda_i}) = 4 \prod_{i=1}^{n} \frac{1+\lambda_i}{2} \cdot \frac{1+\frac{1}{\lambda_i}}{2}$$

$$= 4D(\pi) \le 4M.$$

(b) Let $\alpha_1, \cdots, \alpha_n, > 0$ be eigenvalues of $\frac{1}{2}(\rho_1^\pi + \rho_2^\pi)[(\rho_1^\pi)^{-1} + (\rho_2^\pi)^{-1}]$.

Let $e_1, e_2, \cdots, e_n$ be corresponding eigenvectors, normalized in such a way that

$$<(\frac{\rho_1^\pi+\rho_2^\pi}{2})^{-1} e_j, e_j> = 1, \ j=1,2,\cdots n.$$

Expand $m^\pi = m_1^\pi - m_2^\pi$ in terms of $e_1, \cdots,$ and e_n, i.e.

$$m^\pi = c_1 e_1 + c_2 e_2 + \cdots + c_n e_n.$$

Then $\qquad E(\pi) = c_1^2 + c_2^2 + \cdots + c_n^2$

and $\qquad Q(\pi) = \alpha_1 c_1^2 + \alpha_2 c_2^2 + \cdots + \alpha_n c_n^2.$

Observe that $\frac{1}{2}(\rho_1^\pi+\rho_2^\pi)[(\rho_1^\pi)^{-1}+(\rho_2^\pi)^{-1}] = I + \frac{1}{2}\rho_1^\pi(\rho_2^\pi)^{-1} + \frac{1}{2}[\rho_1^\pi(\rho_2^\pi)^{-1}]^{-1}$

Therefore, $\frac{1}{2}(\rho_1^\pi+\rho_2^\pi)[(\rho_1^\pi)^{-1}+(\rho_2^\pi)^{-1}]$ and $\rho_1^\pi(\rho_2^\pi)^{-1}$ can be simultaneously diagonalized and so we have, for $j=1,\cdots,n$,

$$\begin{aligned}
\alpha_j &= 1 + \frac{1}{2}\lambda_j + \frac{1}{2}\frac{1}{\lambda_j} \\
&= 2\{1 + \frac{(\lambda_j-1)^2}{4\lambda_j}\} \\
&\leq 2 \prod_{k=1}^{n} (1 + \frac{(\lambda_k-1)^2}{4\lambda_k}) \\
&= 2D(\pi) \\
&\leq 2M \qquad\qquad \text{(by Lemma 4.6).}
\end{aligned}$$

Hence $Q(\pi) \leq 2\, M(c_1^2 + \cdots + c_n^2) = 2\, M\, E(\pi)$

$\leq 2\, M^2$ (by Lemma 4.6). #

Theorem 4.2. Suppose $\mu_1 = \mu_1(m_1, \rho_1)$ and $\mu_2 = \mu_2(m_2, \rho_2)$ are two Gaussian measure in $\Omega = \mathbb{R}^\Lambda$. Suppose ρ_1 and ρ_2 are strictly positive definite. Then $H(\mu_1, \mu_2) > 0$ implies that $J(\mu_1, \mu_2) < \infty$.

Proof. This follows from Lemmas (4.2)—(4.7). #

Theorem 4.3. Let $\mu_1 = \mu_1(m_1, \rho_1)$ and $\mu_2 = \mu_2(m_2, \rho_2)$ be two Gaussian measures in $\Omega = \mathbb{R}^\Lambda$. Then μ_1 and μ_2 are either equivalent or orthogonal. They are equivalent if and only if $H(\mu_1, \mu_2) > 0$, which in turn if and only if $J(\mu_1, \mu_2) < \infty$.

Proof. Case 1: Both ρ_1 and ρ_2 are strictly positive definite.

In this case, we have the implication $H(\mu_1, \mu_2) > 0 \Longrightarrow J(\mu_1, \mu_2) < \infty$ by Theorem 4.2. Suppose μ_1 is not orthogonal to μ_2. Then $H(\mu_1, \mu_2) > 0$. Hence $J(\mu_1, \mu_2) < \infty$. Therefore, $\mu_1 \sim \mu_2$ by Property (b) for J. Also, the implication $J(\mu_1, \mu_2) < \infty \Longrightarrow H(\mu_1, \mu_2) > 0$ is obviously true.

Case 2: There is $\pi \in \mathcal{D}$ such that exactly one of $\det \rho_1^\pi$ and $\det \rho_2^\pi$ is zero.

In this case, obviously $\mu_1 \perp \mu_2$.

Case 3: The remaining case.

In this case, we can pick up a maximal set $\Lambda_0 \subset \Lambda$ such that ρ_1 and ρ_2 are strictly positive definite in $\Lambda_0 \times \Lambda_0$. Then we know that either $\mu_1 \sim \mu_2$ or $\mu_1 \perp \mu_2$ in $\mathfrak{B}(\Lambda_0)$. But equivalence and orthogonality with respect to $\mathfrak{B}(\Lambda_0)$ is the same as that with respect to $\mathfrak{B}(\Lambda)$. #

§5. Equivalence and transformation formulas for abstract Wiener measures.

Let (i, H, B) be an abstract Wiener space. As before, p_t denotes the Wiener measure in B with parameter t. For each $x \in B$, define $p_t(x, A) = p_t(A - x)$, $A \in \mathcal{B}(B)$. Thus (i, H, B) carries a family $\{p_t(x, \cdot);\ t > 0,\ x \in B\}$ of Borel measures. Recall that B^* is embedded in H and $(x, y) = \langle x, y \rangle$ whenever $x \in H$ and $y \in B^*$.

Theorem 5.1. $p_t(0, \cdot)$ and $p_t(x, \cdot)$ are either equivalent or orthogonal. They are equivalent if and only if $x \in H$.

Proof. Let $\{e_n\}$ be an orthonormal basis of H with $\{e_n\} \subset B^*$. Define a map Θ from B into $\Omega = \mathbb{R}^{\mathbb{N}}$ by

$$\Theta(x) = ((x, e_1), \cdots, (x, e_n), \cdots).$$

Clearly, $\Theta(H) = \ell_2$. Let $\tilde{B} = \Theta(B)$. Define

$$\| \Theta(x) \| = \| x \|, \qquad \Theta(x) \in \tilde{B}.$$

Let j be the inclusion map from ℓ_2 into $\tilde{B}$. Then $(j, \ell_2, \tilde{B})$ is an abstract Wiener space. Clearly, Θ is unitary from H onto ℓ_2 and isometric from B onto $\tilde{B}$. Observe that if μ_t is the Gauss measure of H with parameter t, then $\mu_t \circ \Theta^{-1}$ is the Gauss measure of ℓ_2 with parameter t. Define

$$\tilde{p}_t(x, A) = p_t(\Theta^{-1}x, \Theta^{-1}(A)), \qquad x \in \tilde{B}, \quad A \in \mathcal{B}(\tilde{B}).$$

Obviously, $\tilde{p}_t(0, \cdot)$ is the σ-additive extension of $\mu_t \circ \Theta^{-1}$ to $\mathcal{B}(\tilde{B})$. For $k=1,\cdots,n,\cdots$, let

$$\upsilon_k(dx) = \frac{1}{\sqrt{2\pi t}} e^{-\frac{x^2}{2t}} dx$$

and

$$\upsilon_k(a_k, dx) = \frac{1}{\sqrt{2\pi t}} e^{-\frac{(x-a_k)^2}{2t}} dx.$$

Define two product measures $\tilde{\upsilon}$ and $\tilde{\upsilon}_a$ in Ω by

$$\tilde{\upsilon} = \upsilon_1 \times \cdots \times \upsilon_k \times \cdots$$

and $\tilde{\upsilon}_a = \upsilon_1(a_1) \times \cdots \times \upsilon_n(a_n) \times \cdots,$

where $a = (a_1, \cdots, a_n, \cdots) \in \Omega$.

By §2 Example 1, $\tilde{\upsilon} \sim \tilde{\upsilon}_a$ if and only if $a \in \ell_2$. Note that $\tilde{\upsilon}$ is the σ-additive extension of $\mu_t \circ \Theta^{-1}$ to the σ-field generated by the cylinder subsets of Ω. Therefore, $\tilde{\upsilon}$ concentrates in $\tilde{B}$ and is the same as $\tilde{p}_t(0, \cdot)$. Moreover, when $a \in \tilde{B}$, $\tilde{\upsilon}_a = \tilde{p}_t(a, \cdot)$. The desired concluison follows immediately. #

<u>Theorem 5.2.</u> If $t \neq s$ then $p_t(0, \cdot)$ and $p_s(x, \cdot)$ are orthogonal for any $x \in B$.

Proof. Use the same argument as in the previous proof except that §2 Example 2 should be used this time. #

Theorem 5.3. $p_t(x, \cdot)$ and $p_s(y, \cdot)$ are either equivalent or orthogonal. They are equivalent if and only if $t = s$ and $x - y \in H$.

Proof. Easy consequence of Theorem 5.1 and Theorem 5.2. #

We now turn to transformation formulas for abstract Wiener measures.

Lemma 5.1. Let $T = I + K$ be a bounded linear operator of B, where I is the identity operator. If $K(B) \subset H$ and $T|_H = I + K|_H$ is an invertible operator of H. Then $T : B \to B$ is also invertible.

Proof. $T^{-1} = I - (T|_H)^{-1}K$. #

Let T be a linear transformation from B into itself as given in Lemma 5.1. Let $p_t \circ T$ denote the Borel measure $p_t \circ T(E) = p_t(T(E))$, $E \in \mathcal{B}(B)$.

Theorem 5.4. Let $T = I + K$ be a linear transformation from B into itself. Assume the following conditions are satisfied:

(a) $K(B) \subset H$,

(b) T is invertible as a map from H into itself, and

(c) $K \in \mathcal{L}_{(1)}(H)$.

Then $p_t \circ T$ and p_t are equivalent and

$$\frac{dp_t \circ T}{dp_t}(x) = e^{-\frac{1}{2t}\{2(Kx, x)+|Kx|^2\}} \det|T|, \quad x \in B.$$

Remarks. By Chapter I Corollary 4.4, (a) implies that $K \in \mathcal{L}_{(2)}(H)$. Segal [42] and Feldman [10] have shown that, under conditions (a) and (b), $p_t \circ T$ and p_t are equivalent and that, under conditions (a), (b) and (c), the Radon-Nikodym derivative $dp_t \circ T / p_t$ can be expressed in the above way.

(Kx, x) is interpreted as a random variable in B as follows. Write $K = S_1^* S_2$, where $S_1, S_2 \in \mathcal{L}_{(2)}(H)$. $\langle S_2h, S_1h\rangle$ as a function in H is uniformly continuous with respect to $\mathcal{T}_m$-topology. Therefore, by Chapter I Theroem 6.2 we have a random variable $\langle S_2h, S_1h\rangle$ defined in B. (Kx, x) is defined to be this random variable. It is easy to see that if P_n is a sequence in $\mathcal{F}$ converging strongly to identity in H such that $P_n(H) \subset B^*$, then (P_nKP_nx, x) as a function in B converges to (Kx, x) in probability as $n \to \infty$.

Proof. It is sufficient to show that for any bounded continuous function f, we have

$$\int_B f(y)p_t(dy) = \int_B f(Tx)g_t(x)p_t(dx),$$

where $$g_t(x) = e^{-\frac{1}{2t}\{2(Kx, x) + |Kx|^2\}} \det|T|.$$

Choose an increasing sequence $\{P_n\}$ of orthogonal projections of H with $\dim P_n(H) = n$ and $P_n(H) \subset B^*$. By (b), $T(P_nH)$ is also n-dimensional. Let Q_n be the orthogonal projection of H onto $T(P_n$ Obviously, $Q_n \to I$ strongly in H as $n \to \infty$. Moreover, T is an isomorphism from P_nH onto Q_nH.

By Chapter I Theorem 6.2 and Theorem 6.3, we have

$$\int_B f(y)p_t(dy) = \lim_{n\to\infty} \int_{Q_nH} f(x)\mu_t(dx),$$

where μ_t is the Gauss measure in H with parameter t.

Consider the transformation $T : P_nH \to Q_nH$. It is easy to see that by making a change of variables, we have

$$\int_{Q_nH} f(x)\mu_t(dx) = a_n \int_{P_nH} f(TP_nx)g_t(P_nx)\mu_t(dx)$$

$$= a_n \int_B f(TP_nx)g_t(P_nx)p_t(dx),$$

where $a_n = \det|I + P_nKP_n| / \det|T|$. Note that $a_n \to 1$ as $n \to \infty$.

By direct computation, we check easily that

$$\int_B g_t(x)p_t(dx) = 1$$

$$\int_B g_t(P_nx)p_t(dx) = a_n^{-1} \to 1 \quad \text{as} \quad n \to \infty.$$

Note that g_t and $g_t \circ P_n$ are positive functions. Moreover, $g_t \circ P_n$ converges to g_t in probability. Choose a subsequence, still denoted by $\{P_n\}$ for convenience, such that $g_t \circ P_n$ converges to g_t almost everywhere (p_t). Therefore, by the following exercise, we have $g_t \circ P_n$ converges to g_t in $L^1(B, p_t)$.

Exercise 35. Let (Ω, μ) be a probability space. Let $1 \leq p < \infty$ $\rho \in L^p(\Omega, \mu)$, $\rho_n \in L^p(\Omega, \mu)$. Suppose $\int_\Omega |\rho_n|^p d\mu \to \int_\Omega |\rho|^p d\mu$

and $\rho_n \to \rho$ a.e. as $n \to \infty$. Then $\rho_n \to \rho$ in $L^p(\Omega, \mu)$.

Recall that f is bounded and $f(TP_n(\cdot))$ converges to $f(T(\cdot))$ in probability as $n \to \infty$. Therefore, by choosing a subsequence, still denoted by $\{P_n\}$, we conclude that

$$\int_B f(y)p_t(dy) = \lim_{n\to\infty} a_n \int_B f(TP_nx)g_t(P_nx)p_t(dx)$$

$$= \int_B f(Tx)g_t(x)p_t(dx). \qquad \#$$

§6. Application of the translation formula Theorem 1.2.

Let B be a Banach space. A function f defined in an open subset U of B is said to be Fréchet differentiable at x if there exists a bounded linear functional x^* such that $|f(x+y)-f(x)-(x^*,y)| = o(\|y\|)$, $y \in B$. It is easy to check that x^* is unique and thus will be denoted by $f'(x)$. f is said to be C^1 if $f'(x)$ exists for each $x \in U$ and the map $x \to f'(x)$ from U into B^* is continuous. In a general Banach space, even separable, the class of bounded C^1-functions is not a very large class of functions, e.g., the works of [1; 47] show that for many separable Banach spaces it is not dense in the space of bounded uniformly continuous functions.

However, if we consider a separable Banach space B as an abstract Wiener space (H, B), then we have a weaker notion of differentiability, with respect to which many nice results can be obtained.

Definition 6.1. Let f be a function from an open subset U of B into a Banach space X. For each $x \in U$, consider the function $g(h)=f(x+h)$, $h \in (U-x) \cap H$. If g is k-times Fréchet differentiable at 0 then we say that f is k-times differentiable in the directions of H at x. (H-differentiable at x). $g^{(j)}(0)$ will be denoted by $D^j f(x)$. $D^j f(x)$ is a j-linear map from $H \times \cdots \times H$ (j factors) into X. We say that f is k-times H-differentiable in U if it is k-times differentiable at each point x of U.

In this section we will prove that any bounded uniformly continuous function can be approximated uniformly by infinitely H-differentiable functions. The approach is as follows. If f is a bounded measurable function, define

$$p_t f(x) = \int_B f(x+y)\, p_t(dy).$$

We will prove that for any bounded uniformly continuous function f, $p_t f$ converges to f uniformly as $t \to 0$. (Theorem 6.1) Then we use the translation formula in Theorem 1.2 to prove that $p_t f$ is infinitely H-differentiable for any bounded measurable function f (Theorem 6.2).

Theorem 6.1. If f is a bounded uniformly continuous function in B, then $p_t f$ converges uniformly to f as $t \to 0$.

Proof. Let $\varepsilon > 0$ be given. Then there exists $\delta > 0$ such that

$$\|x - y\| < \delta \Longrightarrow |f(x) - f(y)| < \varepsilon.$$

Then

$$|p_t f(x) - f(x)| = \left|\int_B (f(x+y) - f(x))\, p_t(dy)\right|$$

$$\leq \int_{\|y\|<\delta} \varepsilon\, p_t(dy) + \int_{\|y\|\geq\delta} |f(x+y) - f(x)|\, p_t(dy$$

$$\leq \varepsilon + 2\|f\|_\infty\, p_t(\|y\| \geq \delta\,),$$

where $||f||_\infty$ is the sup norm of f.

But it is easy to see that $p_{ts}(E) = p_t(\frac{E}{\sqrt{s}})$, $E \in \mathcal{B}(B)$.

Hence

$$p_t(y; ||y|| \geq \delta) = p_1(\frac{y}{\sqrt{t}}; ||y|| \geq \delta)$$

$$= p_1(x; ||x|| \geq \frac{\delta}{\sqrt{t}}) \to 0 \quad \text{as} \quad t \to 0.$$

Hence $||p_t f - f||_\infty \to 0$ as $t \to 0$. #

Theorem 6.2. Let f be a bounded measurable function on B. Then $p_t f$ is infinitely H-differentiable. The first two derivatives are given by

$$\langle D(p_t f)(x), h\rangle = \frac{1}{t}\int_B f(x+y)\langle h,y\rangle p_t(dy), \text{ and}$$

$$\langle D^2(p_t f)(x)h, k\rangle = \frac{1}{t}\int_B f(x+y)\{\frac{\langle h,y\rangle\langle k,y\rangle}{t} - \langle h,k\rangle\} p_t(dy),$$

where $h, k \in H$.

Proof. We want to show that $p_t f(x+h) - p_t f(x) - \frac{1}{t}\int_B f(x+y)\langle h,y\rangle p_t(dy) = o(|h|)$, $h \in H$.

$$p_t f(x+h) = \int_B f(x+h+y) p_t(dy)$$

$$= \int_B f(x+y) p_t(h,dy)$$

$$= \int_B f(x+y) e^{-\frac{1}{2t}|h|^2 + \frac{1}{t}\langle h,y\rangle} p_t(dy) \quad \text{(by Theorem 1.2)}.$$

Let $\quad J(h,y)= e^{-\frac{1}{2t}|h|^2+ \frac{1}{t}< h, y>}$. Then

$$\frac{d}{ds} J(sh, y)= \frac{1}{t}\{<h, y>- s|h|^2\}J(sh, y).$$

Hence

$$p_t f(x+h)-p_t (x)= \int_B \int_0^1 f(x+y)\frac{1}{t}\{<h,y>-s|h|^2\}J(sh,y)ds\, p_t(dy),$$

so, $\quad p_t f(x+h)-p_t f(x)-\frac{1}{t} \int_B f(x+y)<h,y>p_t(dy)$

$$= \int_B f(x+y)\int_0^1 \frac{1}{t}\{<h,y>(J(sh,y)-1)-s|h|^2 J(sh,y)\}ds\, p_t(dy)$$

$$= \int_0^1 \int_B f(x+y)\frac{1}{t}<h,y>(J(sh,y)-1)p_t(dy)ds-\int_0^1\int_B\frac{1}{t}f(x+y)s|h|^2$$

$$p_t(sh,dy)ds.$$

$$\equiv \alpha - \beta, \text{ say.}$$

But,

$$|\alpha|\le \frac{1}{t} \|f\|_\infty \left(\int_B <h,y>^2 p_t(dy)\right)^{\frac{1}{2}}\int_0^1\left(\int_B|J(sh,y)-1|^2 p_t(dy)\right)^{\frac{1}{2}}ds$$

$$= \frac{1}{t} \|f\|_\infty \sqrt{t}|h|\int_0^1(e^{s^2|h|^2/t} - 1)^{\frac{1}{2}} ds,$$

$$|\beta|\le \frac{1}{t}\| f\|_\infty \int_0^1 s|h|^2 ds= \frac{1}{2t}\| f\|_\infty |h|^2.$$

Therefore,

$$|p_t f(x+h) - p_t(x) - \frac{1}{t}\int_B f(x+y)\langle h,y\rangle p_t(dy)|$$

$$\leq |h|\{\frac{1}{\sqrt{t}}||f||_\infty \int_0^1 (e^{s^2|h|^2/t}-1)^{\frac{1}{2}}ds + \frac{1}{2t}||f||_\infty |h|\}$$

$$= o(|h|).$$

The second derivative can be checked using the same technique, so are the higher derivatives, except that the computation is more notationally involved. #

The idea in the previous proof can be used to obtain an integration by parts formula for the abstract Wiener measures. For its proof, see [34].

Theorem 6.3. Let f and g be C_H^1-function from B into Hilbert spaces F and G, respectively. Assume the following conditions hold:

(a) $\int_B (||f(x)||_F)^2(||g(x)||_G)^2 p_t(dx) < \infty$,

(b) there exist constants $r > 0$ and $M < \infty$ such that

$$\int_B ||f(x)||_F \, ||g(x)||_G \, p_t(h,dx) < M \text{ for all } |h| < r, \text{ and}$$

(c) the function

$$\sup_{|h|<r} (||Df(x+h)||_{H,F}||g(x+h)||_G + ||f(x+h)||_F ||Dg(x+h)||_{H,G})$$

is p_t-integrable.

Then for any bounded bilinear map ϕ from $F \times G$ into another Hilbert space K, we have

$$\int_B \phi(f(x),Dg(x)(h))p_t(dx)$$

$$=\int_B \{\frac{1}{t}\langle h,x\rangle\phi(f(x),g(x))-\phi(Df(x)(h),g(x))\}p_t(dx),$$

where $h \in H$.

§7. <u>Comments on Chapter II.</u>

§1. Regarding C as an abstract Wiener space (i, C', C), we see that by Theorem 5.1, w_{x_0} is either equivalent or orthogonal to w. w_{x_0} is equivalent to w if and only if $x_0 \in C'$, i.e. x_0 is absolutely continuous and $x_0' \in L^2[0, 1]$.

§2. Let $\mathcal{B}_n$ be the σ-field in $\mathbb{R}\times\mathbb{R}\times\cdots$ generated by the first n coordinates. Let $X_n = \frac{d\mu_1}{d\upsilon_1}\cdots\cdots\frac{d\mu_n}{d\upsilon_n}$ then X_n is a martingale in $L^1(\mathbb{R}\times\mathbb{R}\times\cdots, \upsilon)$ with respect to $\mathcal{B}_n$. Kakutani's proof can be simplified by using martingale convergence theorem.

§3. This section is based on [46]. Some proofs have been simplified by using the notion of abstract Wiener space.

§4. The proof of Theorem 4.3 is due to Shepp [43].

A general problem discussed by Gihman and Skorokhod is this: Suppose we have a Gaussian measure μ in a topological linear space, then how to characterize the subspace with respect to which μ is quasi-invariant, i.e. μ_x is equivalent to μ iff x belongs to this subspace. For details, see [14].

§5. Our proof of Theorem 5.3 is based on Kakutani's theorem. Gross [19] has another proof using theorems due to Segal [42] and Feldman [10].

Transformation formulas for the classical Wiener space

(i, C', C) can be found in [4; 5]. The proofs there are very tideous.

See [30] for a transformation formula for a non-linear diffeomorphism between two open subsets of B. This transformation formula is used to develop an integration theory on infinite dimensional manifolds.

§6. In Chapter III we will show that $\{p_t\}$ generates a semi-group acting in the Banach space of bounded uniformly continuous functions in B. Obviously, it is a contraction semi-group. Theorem 6.1 tells us that it is strongly continuous. We will see in Chapter III that for a "nice" function f we have

$$\lim_{t \downarrow 0} \frac{1}{t} \{p_t f(x) - f(x)\} = \frac{1}{2} \text{ trace } D^2 f(x).$$

Chapter III. Some results about abstract Wiener space.

In the recent years, there has been a growing research interest in the fields connected with abstract Wiener space and their applications. In this chapter, we merely give a tiny portion of the results obtained so far. Some of theorems here have complete proofs, some have sketchy proofs and, unfortunately, some have no proofs at all. However, we will try to relate these results as close to each other as possible.

§1. Banach space with a Gaussian measure.

Recall that any real separable Banach space can be regarded as an abstract Wiener space (Chapter I Theorem 4.4). More precisely, if B is a given real separable Banach space, then there exists a separable Hilbert space H, which is a dense subspace of B, such that B-norm is measurable on H. Obviously, there are many choices of H. Now, suppose B has a Gaussian measure, can we regard B as an abstract Wiener space (i, H, B) such that this Gaussian measure is the σ-additive extension of the Gauss measure on H?

Definition 1.1. A Borel measure μ in a real separable Banach space B is called Gaussian if for each $y \in B^*$, the random variable (y, x) is normally distributed and there exists $a_\mu \in B$ such that for all $y \in B^*$

$$(y, a_\mu) = \int_B (y, x)\ \mu(dx).$$

a_μ is called the mean of μ.

Theorem 1.1. (Kuelbs [29]) Suppose μ is a Gaussian measure in a real separable Banach space B. Assume that every non-empty open subset of B has positive μ-measure. Then there exists a real separable Hilbert space H such that (i, H, B) is an abstract Wiener space and $\mu = p_1(a_\mu, \cdot)$, where p_1 is the Wiener measure in B with parameter 1 and i is the inclusion map from H into B.

Remarks. (a) It follows from Chapter II Theorem 5.3 that H is uniquely determined by μ.

(b) If μ does not satisfy the above assumption, then the conclusion should be modified. Let B_0 be the smallest closed subspace of B such that $\mu(B_0) = 1$ (B_0 is called the support of μ). Then apply the above theorem to (B_0, μ).

Proof. The theorem is trivial if B is finite dimensional. Hence we assume that dim B = ∞.

Step 1: Since B is separable, we can pick up a countable dense set $\{a_n\}_{n=1}^{\infty}$ in B. For each n, apply the Hahn-Banach theorem to get $z_n \in B^*$ such that $||z_n||_* = 1$ and $(z_n, a_n) = ||a_n||$, where $||\cdot||$ and $||\cdot||_*$ are the norms of B and B*, respectively. Let $\{\lambda_n\}$ be a sequence of positive numbers such that $\sum_{n=1}^{\infty} \lambda_n = 1$. Define, for x and y in B,

$$[x, y] = \sum_{n=1}^{\infty} \lambda_n (z_n, x)(z_n, y).$$

Evidently, [,] is an inner product in B. Let $|x|_0 = \sqrt{[x, x]}$ for x in B. Then

$$|x|_0^2 = \sum_{n=1}^{\infty} \lambda_n (z_n, x)^2 \leq \sum_{n=1}^{\infty} \lambda_n ||x||^2 = ||x||^2.$$

Hence $\qquad |x|_0 \leq ||x|| \qquad$ for all x in B.

Step 2: Let $\tilde{H}$ be the completion of B with respect to $|\cdot|_0$. We show that $\mathcal{B}(\tilde{H}) \cap B = \mathcal{B}(B)$. Note that $\mathcal{B}(\tilde{H}) \cap B \subset \mathcal{B}(B)$ because $|\cdot|_0$

is weaker than $||.||$. Therefore, it is sufficient to show that $\mathcal{B}(B) \subset \mathcal{B}(\tilde{H})$. Obviously, it is sufficient to show that $\{x \in B;\ ||x|| \leq 1\} \in \mathcal{B}(\tilde{H})$.

But, $\{x \in B;\ ||x|| \leq 1\} = \bigcap_{n=1}^{\infty}\{x \in B;\ |(Z_n, x)| \leq 1\}$.

Note that $\tilde{H}^* \subset B^*$. We may and will identify $\tilde{H}^*$ and $\tilde{H}$ canonically. Therefore, we have $B \subset \tilde{H} \subset B^*$. Let $||.||_*$ denote the norm of B^*. Observe that $\tilde{H}$ is dense in B^* because B is dense in $\tilde{H}$. For each n, we choose $\{x_k^{(n)} \in \tilde{H}\}_{k=1}^{\infty}$ such that $||x_k^{(n)}||_* = 1$ and $||x_k^{(n)} - Z_n||_* \leq \frac{1}{k}$. We reorder $\{x_k^{(n)};\ k=1,2,3,\ldots,n=1,2,3,\ldots\}$ as $\{x_1, x_2, \ldots\}$.

Evidently,

$$\{x \in B;\ ||x|| \leq 1\} = \bigcap_{n=1}^{\infty}\{x \in B;\ |[x_n, x]| \leq 1\} \in \mathcal{B}(\tilde{H}).$$

Step 3: Define $\tilde{\mu}(E) = \mu(E \cap B)$, $E \in \mathcal{B}(H)$. Obviously, $\tilde{\mu}$ is a Gaussian measure in $\tilde{H}$ and $m_{\tilde{\mu}} = a_\mu$. Let $S_{\tilde{\mu}}$ be the covariance operator of $\tilde{\mu}$. Then by Chapter I Theorem 2.3, $S_{\tilde{\mu}}$ is an S-operator of $\tilde{H}$. $S_{\tilde{\mu}}$ can be represented by

$$S_{\tilde{\mu}}(x) = \sum_{n=1}^{\infty} c_n [x, e_n]\, e_n,$$

where $c_n > 0$, $\sum_{n=1}^{\infty} c_n < \infty$ and $\{e_n\}$ is an orthonormal basis of $\tilde{H}$.

Let $H = \{x \in \tilde{H};\ x \in \overline{\text{span}}\{e_n\}$ and $\sum_{n=1}^{\infty}\frac{[x, e_n]^2}{c_n} < \infty\}$. Let us write a_μ as a. Define the translation $\tilde{\mu}_a$ of $\tilde{\mu}$ by a by

$$\tilde{\mu}_a(E) = \tilde{\mu}(E + a), \quad E \in \mathcal{B}(\tilde{H}).$$

We show below that $(i, H, \tilde{H})$ is an abstract Wiener space and $\tilde{\mu}_a$ is the Wiener measure in $\tilde{H}$ with parameter 1. For $x, y \in H$, we define

$$\langle x, y\rangle = \sum_{n=1}^{\infty} [x, e_n][y, e_n]/c_n,$$

and let $|x| = \sqrt{\langle x, x\rangle}$ for $x \in H$.

It is easy to see that $[x, y] = \langle\sqrt{S_{\tilde{\mu}}}\, x, \sqrt{S_{\tilde{\mu}}}\, y\rangle$ for all x and y in $\tilde{H}$. Moreover, the same argument as in the proof of Chapter I Theorem 4.3 shows that $\sqrt{S_\mu} \in L_{(2)}(H)$. Therefore, $|\cdot|_0$ is a measurable norm in H. Hence (i, H, $\tilde{H}$) is an abstract Wiener space. Let $x_0' \in H$ be fixed, then

$$\int_{\tilde{H}} e^{i<x_0, y>} d\tilde{\mu}_a(y) = \int_{\tilde{H}} e^{i[S_{\tilde{\mu}}^{-1} x_0, y]} d\tilde{\mu}_a(y)$$

$$= e^{-\frac{1}{2}[S_{\tilde{\mu}} S_{\tilde{\mu}}^{-1} x_0,\ S_{\tilde{\mu}}^{-1} x_0]}$$

$$= e^{-\frac{1}{2}[S_{\tilde{\mu}}^{-1} x_0,\ x_0]}$$

$$= e^{-\frac{1}{2}|x_0|^2}.$$

Therefore, $\tilde{\mu}_a$ is the Wiener measure in $\tilde{H}$ with parameter 1.

Step 4: $H \subset B$ and $||.||$ is a measurable norm in H. Suppose $H \not\subset$ Then we can pick up $x \in H \setminus B$. By Chapter II Theorem 5.3 $\tilde{\mu}$ and $\tilde{\mu}_x$ are equivalent. Hence $\tilde{\mu}(B) = \tilde{\mu}_x(B) = 1$. Note that $B \cap (B + x) = \phi$. Then we have the following contradiction

$$1 = \tilde{\mu}(\tilde{H}) \geq \tilde{\mu}(B) + \tilde{\mu}(B + x) = \tilde{\mu}(B) + \tilde{\mu}_x(B) = 2.$$

By [9, Theorem 2 and Theorem 3] $||\cdot||$ is a measurable norm in H. Hence (i, H, B) is an abstract Wiener space and, clearly, $\mu = p_1(a_\mu, .)$. #

There is another proof of the above theorem by using Feldma theorem [10]. We sketch the approach below. Without loss of generality, we may assume that $a_\mu = 0$. μ_x denotes the translati measure of μ by x, i.e.

$$\mu_x(E) = \mu(E + x), \quad E \in \mathcal{B}(B).$$

Let

$$H = \{x \in B;\ \mu_x \sim \mu\}.$$

For each $x \in H$, define $\langle x^*, y^*\rangle_x = \int_B (x^*, z)(y^*, z)d\mu_x(z)$, $x^*, y^* \in B^*$. It is a consequence of Fernique's theorem [12] that $\int_B ||z||^2 d\mu_x(z) < \infty$. Hence $|x^*|_x^2 \equiv \langle x^*, x^*\rangle_x < \infty$ for all x^* in B^*. Let L_x be the $L_2(B, \mu_x)$-closure of the linear span of B^* and the real-valued constant functions in B. Let

$$T_x : L_0 \longrightarrow L_x, \quad x \in H,$$

be the extension of the identity map $x^* + \alpha \to x^* + \alpha$, where $x^* \in B^*$ and $\alpha \in \mathbb{R}$. By Feldman's theorem T_x is bounded, invertible and $S_x = T_x^* T_x - I \in L_{(2)}(L_x)$. Finally, we define an inner product $\langle\ ,\ \rangle$ in H by

$$\langle x, y\rangle = \langle S_x 1, S_y 1\rangle_0,$$

It can be shown that (i, H, B) is an abstract Wiener space and $\mu = p_1$.

§2. A probabilistic proof of Chapter I Theorem 4.1.

The following lemma is easy to prove.

Lemma 2.1. Let $\{\xi_n\}$ be a sequence of random variables with values in a complete metric space (X, ρ). Suppose $\{\xi_n\}$ is Cauchy in probability, i.e. for any $\varepsilon > 0$, there exists a positive integer N such that

$$\text{Prob}\,\{\omega\ ;\ \rho(\xi_n(\omega), \xi_m(\omega)) > \varepsilon\} < \varepsilon\ , \text{ whenever } n, m \geq N.$$

Then there is a random variable ξ and a subsequence $\{\xi_{n_k}\ ;\ k=1, 2, \ldots\}$ of ξ_n such that $\lim_{k\to\infty} \xi_{n_k} = \xi$ almost surely.

Proof of Chapter I Theorem 4.1 by Kallianpur [27].

It follows from the definition of measurability of $||\cdot||$

that there is an increasing sequence $\{P_n\} \subset \mathcal{F}$, $P_n \to I$ strongly as $n \to \infty$, such that

$$\text{Prob}\ \{\|Px\|^{\sim} > \frac{1}{2^n}\} < \frac{1}{2^n} \quad \text{whenever } P \in F \text{ and } P \perp P_n.$$

Let $\{e_n;\ n=1,2,\dots\}$ be an orthonormal basis of H such that

$$\{e_1, e_2, \dots\dots e_{n_k}\} \text{ is a basis of } P_k(H).$$

Define a sequence of random variables with values in B by:

$$\xi_k(\omega) = \sum_{j=1}^{n_k} n(e_j)(\omega)\ e_j,$$

where $\{n(e_j)\}$ is defined in Chapter I § 4. Note that $\{n(e_j);\ j=1,2,\dots\}$ is an independent sequence of Gaussian variables identically distributed with mean 0 and variance 1. Because

$$\xi_{k+1} - \xi_k = \sum_{j=n_k+1}^{n_{k+1}} n(e_j)e_j = n(P_{k+1}x - P_kx),$$

we have

$$\|\xi_{k+1} - \xi_k\| = \|P_{k+1}x - P_kx\|^{\sim}.$$

But $P_{k+1} - P_k \perp P_k$, hence

$$\text{Prob}\ \{\|\xi_{k+1} - \xi_k\| > \frac{1}{2^k}\} < \frac{1}{2^k}.$$

Therefore, $\{\xi_k\}$ is Cauchy in probability. By the above lemma there is a random variable ξ and a subsequence of ξ_k, still denoted by ξ_k for convenience, such that $\xi_k \to \xi$ a.s.

Let ν be the distribution of ξ, i.e.

$$\nu(E) = \text{Prob}\{\xi^{-1}(E)\}, \qquad E \in \mathcal{B}(B).$$

Clearly, ν is a Borel measure in B. We will finish the proof by showing that ν is the extension of the Gauss measure in H. To do this, it is sufficient to show that for $z \in B^*$ $(B^* \subset H \subset B)$, we have

$$\int_B e^{i(z,x)}\nu(dx) = e^{-\frac{1}{2}|z|^2}.$$

But
$$\int_B e^{i(z,x)}\nu(dx) = \int_\Omega e^{i(z,\,\xi(\omega))}P(d\omega)$$
$$= \lim_{k\to\infty} \int_\Omega e^{i(z,\,\xi_k(\omega))}P(d\omega)$$
$$= \lim_{k\to\infty} E(e^{i(z,\sum_{j=1}^{n_k} n(e_j)e_j)})$$
$$= \lim_{k\to\infty} \prod_{j=1}^{n_k} E(e^{i(z,\,e_j)n(e_j)})$$
$$= \lim_{k\to\infty} \prod_{j=1}^{n_k} e^{-\frac{1}{2}<z,\,e_j>^2}$$
$$= e^{-\frac{1}{2}\sum_{j=1}^{\infty}<z,\,e_j>^2}$$
$$= e^{-\frac{1}{2}|z|^2}.$$

#

§3. Integrability of $e^{\alpha||x||^2}$ and $e^{\beta||x||}$.

Let (i, H, B) be an abstract Wiener space and $\{p_t;\ t>0\}$ be its Wiener measures. Let Ω denote the space of continuous functions ω on $[0,\infty)$ with values in B and $\omega(0)=0$. Then there exists a unique probability measure P on the σ-field generated by the coordinate functions $\omega \to \omega(t)$ for $t>0$, such that if $0=t_0<t_1<\ldots\ldots<t_n$ then $\omega(t_j)-\omega(t_{j-1})$, $j=1,2,\ldots\ldots n$, are independent and the jth one is distributed in B according to $p_{t_j-t_{j-1}}$. Define $W(t): \Omega \to B$ by $W(t)(\omega)=\omega(t)$. W(t) is called a <u>Wiener process</u> in B starting at the origin. Observe that

$$\int_B e^{\alpha||x||^2}p_1(dx) = E[e^{\alpha||W(1)||^2}].$$

<u>Theorem 3.1.</u> (Fernique [12]). There exists $\alpha>0$ such that

$\int_B e^{\alpha||x||^2} p_1(dx) < \infty.$

Proof. Let $X = W(1)$ and $Y = W(2) - W(1)$. Then X and Y are independent and distributed according to p_1. Moreover $\frac{X+Y}{\sqrt{2}}$ and $\frac{X-Y}{\sqrt{2}}$ are independent and distributed according to p_1. Suppose $t > s$, we have

$$P(||W(1)|| \leq s)P(||W(1)|| > t)$$

$$=P(\frac{||X+Y||}{\sqrt{2}} \leq s)P(\frac{||X-Y||}{\sqrt{2}} > t)$$

$$= P(\frac{||X+Y||}{\sqrt{2}} \leq s \text{ and } \frac{||X-Y||}{\sqrt{2}} > t)$$

$$\leq P(\left|||X|| - ||Y||\right| \leq \sqrt{2}\, s \text{ and } ||X||+||Y||>\sqrt{2}$$

Observe that in the xy-plane the region $|x-y| \leq \sqrt{2}\, s$, $x+y > \sqrt{2}\, t$ in the first quadrant is given by

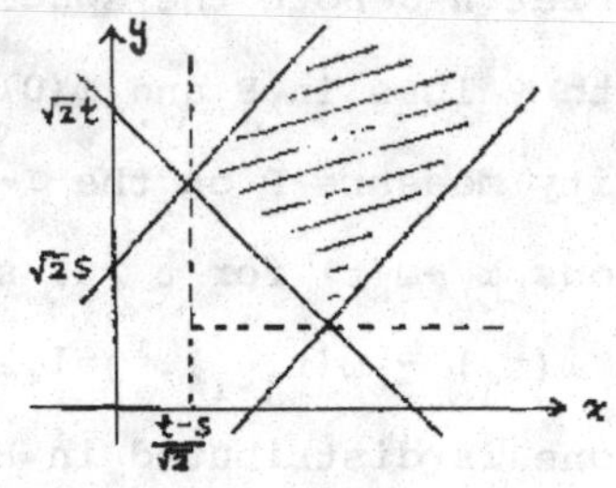

This region is contained in the region $x > \frac{t-s}{\sqrt{2}}$ and $y > \frac{t-s}{\sqrt{2}}$. Therefore,

$$P(||W(1)|| \leq s)P(||W(1)|| > t) \leq P(||X||>\frac{t-s}{\sqrt{2}} \text{ and } ||Y||>\frac{t-s}{\sqrt{2}}$$

$$= P(||X||> \frac{t-s}{\sqrt{2}})P(||Y|| > \frac{t-s}{\sqrt{2}})$$

$$= P(||W(1)|| > \frac{t-s}{\sqrt{2}})^2.$$

Hence we have obtained the relation,

$$P(||W(1)|| \leq s)P(||W(1)|| > t) \leq P(||W(1)|| > \frac{t-s}{\sqrt{2}})^2.$$

Define a sequence of numbers t_n inductively by

$$t_0 = s > 0, \quad t_{n+1} = s + \sqrt{2}\, t_n.$$

Obviously, $t_n = (1+\sqrt{2}+.....+(\sqrt{2})^n)s = \frac{(\sqrt{2})^{n+1} - 1}{\sqrt{2} - 1} s$

$$=((\sqrt{2})^{n+1} - 1)(\sqrt{2} + 1)s.$$

Let $\alpha_n = \frac{P(||W(1)|| > t_n)}{P(||W(1)|| \leq s)}, \quad n \geq 0.$

Then the above relation gives

$$\alpha_{n+1} \leq \alpha_n^2 .$$

Therefore, $\alpha_n \leq \alpha_0^{2^n} = e^{2^n \log \alpha_0}.$

Hence

$$P(||W(1)|| > ((\sqrt{2})^{n+1} - 1)(\sqrt{2} + 1)s)$$

$$= \alpha_n P(||W(1)|| \leq s)$$

$$\leq P(||W(1)|| \leq s)e^{2^n \log \alpha_0}.$$

Let $u = (\sqrt{2})^{n+4} s$. Then $u > ((\sqrt{2})^{n+1} - 1)(\sqrt{2} + 1)s$ and we have

$$P(||W(1)|| > u) \leq P(||W(1)|| \leq s)\exp[\frac{u^2}{16s^2}\log\alpha_0].$$

Obviously, the above relation holds for all $u \geq 4s$.

Choose s so large that $\alpha_0 < 1$. Let

$$a = -\log\alpha_0/16s^2 \quad \text{and} \quad b = P(||W(1)|| \leq s).$$

Choose a positive integer $N \geq 4s$ then we have

$$P(||W(1)|| > u) \leq be^{-au^2}, \quad u \geq N.$$

Therefore

$$\int_{||x||>N} e^{\alpha||x||^2} p_1(dx) = \sum_{k=N}^{\infty} \int_{k<||x||\leq k+1} e^{\alpha||x||^2} p_1(dx)$$

$$\leq \sum_{k=N}^{\infty} e^{\alpha(k+1)^2} P(||W(1)|| > k)$$

$$\leq b \sum_{k=N}^{\infty} e^{\alpha(k+1)^2} e^{-ak^2}.$$

Obviously, we can choose α so that $0 < \alpha < a$. Then the last series above is convergent and we have

$$\int_{||x||>N} e^{\alpha||x||^2} p_1(dx) < \infty,$$

which gives immediately the conclusion of the theorem. #

Theorem 3.2 (Skorokhod [44]) There exists $\beta > 0$ such that $\int_B e^{\beta||x||} p_1(dx) < \infty$.

Remark. Obviously, this theorem is a consequence of Theorem 3.1. Skorokhod obtained independently this weaker result in the same time as Fernique did in 1970. Since Skorokhod's proof is more probabilistic in nature, we present the original proof here.

Proof. Take $0 < \varepsilon < 1$. Then there exists $\delta > 0$ such that

$$P\{\sup_{0\leq t\leq\delta} ||W(t)|| > 1\} < \varepsilon .$$

Let τ_n be the exit time of $W(t)$ from the sphere $\{x \in B; ||x|| < n\}$.

Then,

$$\{\omega;\ \sup_{0\leq t\leq\delta} ||W(t)(\omega)|| \geq n\} \subset \{\omega; \tau_{n-1}(\omega) < \delta \text{ and } \sup_{\tau_{n-1}\leq t\leq\delta} ||W(t)(\omega) - W(\tau_{n-1})|| > 1\}.$$

Therefore, by the independence of τ_{n-1} and $\sup_{\tau_{n-1}\leq t\leq\delta} ||W(t)-W(\tau_{n-1})||$,

$$P\{\sup_{0\leq t\leq\delta} ||W(t)|| \geq n\} \leq P\{\tau_{n-1} < \delta\}\, P\{\sup_{\tau_{n-1}\leq t\leq\delta} ||W(t)-W(\tau_{n-1})|| > 1\}.$$

But, since $W(t)- W(\tau_{n-1})$, $t \geq \tau_{n-1}$, is also a Wiener process, we have

$$P\{\sup_{\tau_{n-1}\leq t\leq\delta} ||W(t)- W(\tau_{n-1})|| > 1\}= P\{\sup_{0\leq t\leq\delta} ||W(t)|| > 1\}.$$

Hence,

$$P\{\sup_{0\le t\le\delta} ||W(t)|| \ge n\} \le \varepsilon P\{\tau_{n-1} < \delta\}$$

$$\le \varepsilon P\{\sup_{0\le t\le\delta} ||W(t)|| \ge n-1\}.$$

This implies that

$$P\{\sup_{0\le t\le\delta} ||W(t)|| \ge n\} \le \varepsilon^n.$$

Note that for any $a > 0$, $\sqrt{a}W(\frac{t}{a})$ is also a Wiener process. Thus we have

$$P\{||W(1)|| \ge \frac{n}{\sqrt{\delta}}\} = P\{||W(\delta)|| \ge n\}$$

$$\le P\{\sup_{0\le t\le\delta} ||W(t)|| \ge n\}$$

$$\le \varepsilon^n.$$

Therefore,

$$\int_B e^{\beta||x||} p_1(dx) = E\{e^{\beta||W(1)||}\}$$

$$= \sum_{n=0}^{\infty} E\{e^{\beta||W(1)||}; \frac{n}{\sqrt{\delta}} \le ||W(1)|| \le \frac{n+1}{\sqrt{\delta}}\}$$

$$\le e^{\beta\frac{1}{\sqrt{\delta}}} + \sum_{n=1}^{\infty} e^{\beta\frac{n+1}{\sqrt{\delta}}} P\{||W(1)|| \ge \frac{n}{\sqrt{\delta}}\}$$

$$\le e^{\beta\frac{1}{\sqrt{\delta}}} + \sum_{n=1}^{\infty} e^{\beta\frac{n+1}{\sqrt{\delta}}} \varepsilon^n$$

$$= e^{\beta\frac{1}{\sqrt{\delta}}} + e^{\beta\frac{1}{\sqrt{\delta}}} \sum_{n=1}^{\infty} (\varepsilon e^{\frac{\beta}{\sqrt{\delta}}})^n.$$

Choose β so that $0 < \beta < -\sqrt{\delta}\log\varepsilon$. Then the last series above is convergent and so $\int_B e^{\beta||x||} p_1(dx) < \infty$. #

§4. Potential theory.

Consider the Poisson equation $\frac{1}{2}\Delta u = -f$ in $\mathbb{R}^n$, where $\Delta = \partial^2/\partial x_1^2 + \ldots + \partial^2/\partial x_n^2$. Recall that a solution of this equation is given by the potential of f, i.e.

$$u(x) = \int_{R^n} f(x-y) G^{(n)}(dy),$$

where $G^{(n)}(dy)$ is the Green measure in $\mathbb{R}^n$,

$$G^{(n)}(dy) = \frac{2}{(n-2)\,\omega_n} |y|^{2-n} dy, \quad n \geq 3 .$$

Here $\omega_n = 2(\sqrt{\pi})^n/\Gamma(\frac{n}{2})$ is the surface area of the unit sphere of $\mathbb{R}^n$.

A direct generalization of $G^{(n)}(dy)$ to infinite dimensional space seems to be impossible from the above formula. However, it is easy to check that

$$G^{(n)}(dy) = \int_0^\infty p_t^{(n)}(dy)\,dt,$$

where $\quad p_t^{(n)}(dy) = (2\pi t)^{-n/2} e^{-|y|^2/2t} dy.$

This suggests that in an abstract Wiener space (i, H, B) we can define the Green measure G(dy) by

$$G(dy) = \int_0^\infty p_t(dy)\,dt$$

and define the potential of f by

$$Gf(x) = \int_B f(x - y)G(dy).$$

However, a new kind of regularity problem arises in ∞-dimensional case, i.e., whether the second derivative of Gf is a Hilbert-Schmidt operator, or trace class operator. We will answer this type of question later on in this section. We study first some properties of the Wiener measure p_t. If f is a bounded measurable function, define

$$(p_t f)(x) = \int_B f(x + y)p_t(dy), \ t > 0,$$

and $p_0 f = f$.

Most of the following theorems have been proved in [19]. However, many of the original proofs have been simplified here.

Theorem 4.1. $\{p_t; t \geq 0\}$ form a strongly continuous contraction semigroup on the Banach space $\mathfrak{A}$ of bounded uniformly continuous complex valued functions on B.

Proof. Clearly, $p_t f \in \mathfrak{A}$ whenever $f \in \mathfrak{A}$. By Chapter II §7 comments on §6, we need only to show that $p_t * p_s = p_{t+s}$, where * denotes the convolution, i.e.

$$p_t * p_s(E) = \int_B p_s(E - x)p_t(dx), \ E \in \mathcal{B}(B).$$

Let E be a cylinder set, say $E = \{y \in B; ((y,e_1)\ldots,(y,e_k)) \in F\}$, where $\{e_j; j=1,..,k\} \subset B^*$ and are orthonormal in H. Define a projection Q of B by

$$Qy = (y, e_1)e_1 + \dots + (y, e_k)e_k, \; y \in B.$$

It is easy to see that $E - x = E - Qx$ for any $x \in B$. Therefore, $p_s(E-x) = p_s(E-Qx) = \mu_s(E-Qx)$, where μ_s is the Gauss measure in the span K of $\{e_j;\ j=1,\dots,k\}$ in H with parameter s. Moreover, $\mu_s(E-Qx)$ is a cylinder function based in K. Hence

$$\int_B \mu_s(E - Qx)p_t(dx) = \int_K \mu_s(E - Qx)\mu_t(dx).$$

But $\mu_t * \mu_s = \mu_{t+s}$ in K. Hence

$$\int_B \mu_s(E - Qx)p_t(dx) = \mu_{t+s}(E)$$

$$= p_{t+s}(E)$$

Thus we have shown that for any cylinder set E

$$\int_B p_s(E - x)p_t(dx) = p_{t+s}(E).$$

It is easy to see that the above relation holds for any Borel subset E of B. #

We can ask now the question: what is the infinitesmal generator of $\{p_t;\ t \geq 0\}$?

There are two relevant senses of differentiability for functions defined in an open subset V of B. Let f be a function from V into a Banach space X. f is said to be <u>Fréchet differentiable</u> at $x \in V$ if there exists $K \in L(B, X)$ such that $||f(x+y) - f(x) - Ky||_X = o(||y||)$, $y \in B$. K is easily seen to be unique and will be denoted by $f'(x)$. f is said to be $\underline{C}^1$ if $f'(x)$ exists for each x in V and f' is continuous from V into the Banach space $L(B, X)$. Higher derivatives $f^{(k)}(x)$ and C^k, $k \geq 2$ are defined similarly. Another kind of differentiability is <u>H-differentiability</u> defined in Chapter II Definition 6.1. $D^k f(x)$, $k \geq 1$, will denote the k-th H-derivative of f at x. Obviously, if f is k times Fréchet differentiable then it is also k times H-differentiable and $D^j f$ is the restriction of $f^{(j)}$ to $H \times \ldots \times H$ (j factors) $1 \leq j \leq k$. However, the converse is not true. In fact, an H-differentiable function may not every be continous with respect to $||.||$-topology in B.

<u>Example 1.</u> Let $h \in H$. Define $f(x) = \langle x, h\rangle$ if $x \in H$ and $f(x)=0$ if $x \in B \setminus H$. Then f is infinitely H-differentiable. If $x \in H$ then $Df(x) = h$ and $D^j f(x) = 0$ for $j \geq 2$. If $x \in B \setminus H$ then $D^j f(x) = 0$ for $j \geq 1$.

<u>Theorem 4.2</u>. Let $u \in \mathfrak{A}$ be such that u' and u" exist, u" is bounded and is uniformly $||.||$-continuous from B into $L(B, B^*)$ (with the weak operator topology). Then u is in the domain of the infinitesmal generator G of the semi-group $\{p_t;\ t \geq 0\}$ and

$$Gu(x) = \frac{1}{2} \text{ trace } [D^2 u(x)].$$

Proof. For x and y in B, we have

$$u(x+y)-u(x)=(u'(x), y)+\int_0^1(1-s)(u''(x+sy)y, y)ds.$$

Since $p_t(-E)= p_t(E)$ for all $E \in \mathcal{B}(B)$, we have

$\int_B(u'(x), y)p_t(dy) = 0$ for all x in B. Therefore,

$$\frac{1}{t}(p_t u(x)-u(x))=\int_0^1(1-s)\int_B(u''(x+sy)y, y)\frac{1}{t}p_t(dy)ds$$

$$=\int_0^1(1-s)\int_B(u''(x+s\sqrt{t}\,y)y, y)p_1(dy)ds.$$

Here we have used the fact that $p_t(E)= p_1(\frac{E}{\sqrt{t}})$ for any Borel subset E of B.

On the other hand, it is easy to check that $\int_B(Tx, x)p_1(dx)$ $=\text{trace } T$ for $T \in \mathcal{L}_{(1)}(H)$, (cf. Exercise 25). Therefore,

$$\text{trace } D^2u(x) = \int_B(u''(x)y, y)p_1(dy).$$

Hence,

$$\left|\frac{1}{t}(p_t u(x)-u(x)-\frac{1}{2}\text{trace } D^2u(x)\right|$$

$$=\left|\int_0^1(1-s)\int_B([u''(x+s\sqrt{t}\,y)-u''(x)]y, y)p_1(dy)ds\right|$$

$$\leq\int_0^1(1-s)\int_B\theta_t(s, y)p_1(dy)ds,$$

where $\theta_t(s, y)=\sup_{x\in B}|([u''(x+s\sqrt{t}\,y)-u''(x)]y, y)|$. Note that if

$\{x_n\}$ is a countable dense subset of B, then by the continuity of u" we have

$$\theta_t(s, y)= \sup_n |([u''(x_n+s\sqrt{t}y)- u''(x_n)]y, y)|.$$

Therefore, $\theta_t(s, y)$ is a measurable function of s and y. By the uniform continuity of u", we ahve

$$\lim_{t\downarrow 0} \theta_t(s, y) = 0 \quad \text{for each s and y.}$$

Moreover $\theta_t(s, y) \leq \beta||y||^2$, where $\beta = \sup_{x\in B}||u''(x)||_{B,B^*} < \infty$.

By the Lebesgue dominated convergence theorem

$$\lim_{t\downarrow 0} \frac{1}{t}(p_t u(x) - u(x)) = \frac{1}{2} \text{ trace } D^2u(x) , \quad \text{uniformly in x.}$$

This shows that trace $D^2u \in \mathcal{O}\!\mathcal{L}$ and

$$Gu(x) = \frac{1}{2} \text{ trace } D^2u(x) \qquad \#$$

Let W(t) be a Wiener process in B starting at the origin i.e W(t) has continuous sample paths, W(0)= 0 a.s., W(.) has independent increments and W(t)- W(s) is distributed according to p_{t-s}, $t > s$. If $x \in B$, x + W(t) is a Wiener process starting at x. Let τ_x^V denote the first exit time for x + W(t) from the open set V, i.e.,

$$\tau_x^V = \inf\{t \geq 0; x + W(t) \notin V\}.$$

If $V = \{y; ||y-x||< r\}$, τ_x^V will be denoted by $\tau_x^{(r)}$. It is easy to check that $E[\tau_x^{(r)}] < \infty$.

Definition 4.1. Let f be a measurable function defined in a neighborhood of x in B. The generalized Laplacian $\Delta f(x)$ of f at x is defined by

$$\Delta f(x) = 2 \lim_{r \downarrow 0} \frac{E[f(x+W(\tau_x^{(r)}))] - f(x)}{E[\tau_x^{(r)}]}$$

when it exists.

Definition 4.2. The closed convex hull determined by x and a ball not containing x is called a cone with vertex x. An open set V in B is called strongly regular at $x \in \partial V$ (∂V is the boundary of V) if there is a cone K with vertex x and $V \cap K = \phi$. V is strongly regular if it is strongly regular at each point of ∂V.

Theorem 4.3. Let $V \subset B$ be a strongly regular open set. Let ρ be a bounded continuous function on ∂V. Then the function

$$u(x) = E[\rho(x + W(\tau_x^V)); \tau_x^V < \infty]$$

satisfies $\Delta u = 0$ and $u(x) = \rho(x)$ for x in ∂V. That is, u solves the Dirichlet problem for V and ρ.

In fact, it is easy to show that $u(x) = E[u(x + W(\tau_x^{(r)}))]$ and $u = \rho$ on ∂V.

Lemma 4.1. Let $f(x) = \frac{1}{2}(Ax, x)$, $A \in L(B, B^*)$. Then $\Delta f(x) =$ trace

$(A|_H)$ for all x in B.

This lemma can be proved by stochactic integrals (see next section).

Theorem 4.4. Let u be a twice Fréchet differentiable function in an open subset V of B with bounded second derivative. Let u" from V into $L(B, B^*)$ (with the weak operator topology) be continuous at x. Then

$$\Delta u(x) = \text{trace } D^2 u(x).$$

Proof. Let π_r be the surface measure in $S_r = \{x \in B; \|x\| = r\}$ determined by the Wiener process W, i.e.

$$\pi_r(A) = \text{prob}\{W(\tau_0^{(r)}) \in A\}, \quad A \in \mathcal{B}(S_r).$$

Then we have

$$\Delta u(x) = \frac{2}{c} \lim_{r \downarrow 0} \frac{1}{r^2} \int_{S_r} (u(x+y) - u(x)) \pi_r(dy),$$

where $c = E[\tau_0^{(1)}]$. Note that π_r is even and $\pi_r(A) = \pi_1(\frac{A}{r})$ for $A \in \mathcal{B}(S_r)$. Therefore,

$$\Delta u(x) = \frac{2}{c} \lim_{r \downarrow 0} \int_0^1 (1-s) \int_{S_1} (u''(x+rsy)y, y) \pi_1(dy) ds.$$

By the Lebesgue dominated convergence theorem,

$$\Delta u(x) = \frac{1}{c} \int_{S_1} (u''(x)y, y) \pi_1(dy).$$

On the other hand, from the above lemma we have

$$\begin{aligned}\operatorname{trace}(A|_H) &= \Delta f(0)\\ &= \frac{2}{c}\lim_{r\downarrow 0}\frac{1}{r^2}\int_{S_r}\frac{1}{2}(Ay, y)\pi_r(dy)\\ &= \frac{1}{c}\int_{S_1}(Ay, y)\pi_1(dy).\end{aligned}$$

Therefore,

$$\Delta u(x) = \operatorname{trace} D^2u(x). \qquad \#$$

Theorem 4.5. Let u be a twice H-differentiable function in an open subset U of B such that D^2u is continuous from U into $\mathcal{L}_{(1)}(H)$. Then

$$\Delta u(x) = \operatorname{trace} D^2u(x), \quad x \in U.$$

This theorem can be proved by stochactic integrals (see §5).

Lemma 4.2. Let g be a bounded C_H^1 function in B and $\sup_{x\in B}|Dg(x)| < \infty$. Then p_tg is also C_H^1 and

$$\langle D(p_tg)(x), h\rangle = \int_B \langle Dg(x+y), h\rangle p_t(dy), \quad h \in H.$$

Lemma 4.3. Let f be a bounded Lip-1 function in B. Then p_tf is twice H-differentiable and

$$\langle D^2(p_tf)(x)h, k\rangle = \frac{1}{at}\int_B \langle Dp_{bt}f(x+y), h\rangle\langle y, k\rangle p_{at}(dy),$$

where a > 0, b > 0 and a + b = 1.

Proof. Write $p_t f(x) = \int_B p_{bt} f(x+y) p_{at}(dy)$. Then apply Lemma 4.2 and Chapter II Theorem 6.2. #

<u>Definition 4.3</u>. A <u>test operator</u> is a bounded operator T of B with finite dimensional range contained in B*.

Remark. If T is a test operator, we can regard T as a bounded operator of H. We will use $\tilde{T}$ to denote T when we view it as an operator of H.

<u>Lemma 4.4</u>. Let f be as in Lemma 4.3 and T a test operator. Then

$$\text{trace } \tilde{T} D^2 p_t f(x) = \frac{1}{at} \int_B \langle D p_{bt} f(x+y), Ty \rangle \, p_{at}(dy),$$

where $a > 0$, $b > 0$ and $a+b=1$.

<u>Theorem 4.6</u>. Let f be a bounded Lip-1 function in B. Then

(a) $D^2 p_t f(x) \in \mathcal{L}_{(1)}(H)$ for all x in B,

(b) For each $c > 0$, the map $(t, x) \mapsto D^2 p_t f(x)$ is uniformly continuous from $[c, \infty) \times B$ into $\mathcal{L}_{(1)}(H)$,

(c) $v(t, x) \equiv p_t f(x)$ is uniformly continuous in $[0, \infty) \times B$.

(d) For each $t > 0$, $\partial v / \partial t$ exists uniformly in x and $\partial v / \partial t = \frac{1}{2}$ trace $D^2 v(t, x)$.

Remark. By Theorem 4.1 $\lim_{t \downarrow 0} v(t, x) = f(x)$ uniformly in x. Hence $p_t f(x)$ solves the heat equation $\partial u / \partial t = \frac{1}{2}$ trace $D^2 u(t, x)$

with initial conditon f, which is assumed to be bounded and Lipschitzian.

Proof.

(a). Let β be a Lipschitzian constant of f, i.e.

$$|f(x) - f(y)| \leq \beta||x - y||, \quad x,y \in B.$$

Then for any $t > 0$, we also have

$$|p_t f(x) - p_t f(y)| \leq \beta||x - y||, \quad x, y \in B.$$

Hence, for $h \in H$,

$$|\langle Dp_t f(x), h\rangle| = \lim_{\varepsilon\to 0} \frac{1}{\varepsilon}|p_t f(x+\varepsilon h) - p_t f(x)| \leq \beta||h||.$$

Therefore, by Exercise 19 $Dp_t f(x) \in B^*$ and $||Dp_t f(x)||_* \leq \beta$ for all $t > 0$ and $x \in B$. Here $||\cdot||_*$ denotes B^*-norm.

Apply Lemma 4.4 to obtain

$$|\text{trace } \tilde{T}D^2 p_t f(x)| \leq \frac{1}{at} \beta\int_B ||Ty|| p_{at}(dy)$$

$$\leq \frac{1}{at}\beta||\tilde{T}||\int_B ||y|| p_{at}(dy) \quad \text{(by Chapter I Theorem 4.5).}$$

$$= (at)^{-1/2} \beta\, ||\tilde{T}||\int_B ||y|| p_1(dy).$$

This holds for any $0 < a < 1$. Therefore

$$|\text{trace } \tilde{T}D^2p_tf(x)| \leq \beta t^{-1/2} ||\tilde{T}|| \int_B ||y|| p_1(dy).$$

By the same argument as in Chapter I Theorem 4.6, we conclude that for all x in B, $D^2p_tf(x) \in \mathcal{L}_{(1)}(H)$ and

$$||D^2p_tf(x)||_1 \leq \beta t^{-1/2} \int_B ||y|| p_1(dy).$$

(b). Step 1: By Chapter II Theorem 6.2

$$\langle Dp_tf(x), h\rangle = \frac{1}{\sqrt{t}} \int_B f(x+\sqrt{t}y) \langle y, h\rangle p_1(dy), \quad h \in H.$$

Using the assumption that f is bounded and Lipschitzian, we see that there is a constant $\alpha > 0$ such that whenever $t, s \geq c > 0$ and $x, x' \in B$, we have

$$|Dp_tf(x) - Dp_sf(x')| \leq \alpha[|\sqrt{t}-\sqrt{s}| + ||x-x'||].$$

From Lemma 4.4, we have

$$\text{trace } \tilde{T}D^2p_tf(x) = \frac{1}{\sqrt{at}} \int_B \langle Dp_{bt}f(x+\sqrt{at}y), Ty\rangle p_1(dy).$$

Hence,

$$|\text{trace } \tilde{T}D^2p_tf(x) - \text{trace } \tilde{T}D^2p_sf(x')|$$

$$\leq \left| \frac{1}{\sqrt{at}} - \frac{1}{\sqrt{as}} \right| \beta \int_B ||Ty|| p_1(dy)$$

$$+ \frac{1}{\sqrt{as}} \int_B \alpha[|\sqrt{bs}-\sqrt{bt}| + ||x-x'|| + |\sqrt{as}-\sqrt{at}|\, ||y||] \, |Ty| p_1(dy).$$

Since b < 1 and we may let a → 1, we have

$$|\text{trace } \tilde{T}(D^2p_t f(x) - D^2p_s f(x'))|$$

$$\leq \beta\left|\frac{1}{\sqrt{t}} - \frac{1}{\sqrt{s}}\right| \left(\int_B ||y|| p_1(dy)\right) ||\tilde{T}||$$

$$+ \frac{\alpha}{\sqrt{s}} \int_B [|\sqrt{s}-\sqrt{t}| + ||x-x'|| + |\sqrt{s}-\sqrt{t}|\,||y||]\,|Ty|\,p_1(dy).$$

Recall that

$$\int_B |Ty| p_1(dy) \leq \left(\int_B |Ty|^2 p_1(dy)\right)^{\frac{1}{2}} = ||\tilde{T}||_2 \text{ (see Exercise 25)},$$

and note that $||\tilde{T}|| \leq ||\tilde{T}||_2$. Therefore, we get

$$||D^2p_t f(x) - D^2p_s f(x')||_2$$

$$\leq \beta\left|\frac{1}{\sqrt{t}} - \frac{1}{\sqrt{s}}\right| \int_B ||y|| p_1(dy)$$

$$+ \frac{\alpha}{\sqrt{s}} \{|\sqrt{s}-\sqrt{t}| + ||x-x'|| + |\sqrt{s}-\sqrt{t}| \left(\int_B ||y||^2 p_1(dy)\right)^{\frac{1}{2}}\}.$$

This shows that $D^2p_{(.)}f(.) : [c, \infty) \times B \to L_{(2)}(H)$ is uniformly continuous for any $c > 0$.

Step 2: Let Q_n and $(B_0, ||.||_0)$ be as in Chapter I Corollary 4.2. By the Uniform Boundedness Principle $\sup_n ||Q_n||_{B_0, B_0} = M < \infty$. Hence

$$||Q_n y||_0 \leq M ||y||_0 \quad \text{for all } n.$$

Note that $\int_{B_0} ||y||_0 \tilde{p}_1(dy) < \infty$, where $\tilde{p}_1$ is the Wiener measure in B_0 with parameter 1. So, by the Lebesgue dominated convergence theorem

$$\lim_{n\to\infty} \int_{B_0} ||Q_n y - y||_0 \tilde{p}_1(dy) = 0.$$

Step 3: By step 1, $D^2 p_{(.)} f(.)$ is uniformly continuous from $[c, \infty) \times B$ into $L_{(2)}(H)$, hence so is $Q_n D^2 p_{(.)} f(.)$ for each n. Note that for any $A \in L_{(2)}(H)$

$$||Q_n A||_1 \leq [\dim Q_n(H)]^{1/2} ||A||_2.$$

Therefore, for each n the map $Q_n D^2 p_{(.)} f(.)$ is uniformly continuous from $[c, \infty) \times B$ into $L_{(1)}(H)$. Thus to finish the proof of (b) we need only to show that

$$\lim_{n\to\infty} ||Q_n D^2 p_t f(x) - D^2 p_t f(x)||_1 = 0$$

uniformly for $t > c$ and $x \in B$.

From Lemma 4.4 we have

$$\text{trace } \tilde{T} D^2 p_t f(x) = \frac{1}{\sqrt{at}} \int_B \langle Dp_{bt} f(x + \sqrt{at}y), Ty \rangle p_1(dy)$$

and

$$\text{trace } \tilde{T} Q_n D^2 p_t f(x) = \frac{1}{\sqrt{at}} \int_B \langle Dp_{bt} f(x + \sqrt{at}y), TQ_n y \rangle p_1(dy).$$

Also, from the proof of (a) we have

$\|Dp_t f(x)\|_* \leq \beta$ for all $t > 0$ and x in B.

Hence

$$|\text{trace } \tilde{T}(Q_n D^2 p_t f(x) - D^2 p_t f(x)| \leq \frac{\beta}{\sqrt{at}} \int_B \|T(Q_n y - y)\| \, p_1(dy)$$

$$\leq \frac{\beta}{\sqrt{at}} \int_B \| Q_n y - y \| p_1(dy) \ \|\tilde{T}\|$$

Thus

$$\|Q_n D^2 p_t f(x) - D^2 p_t f(x)\|_1 \leq \frac{\beta}{\sqrt{at}} \int_B \|Q_n y - y\| \, p_1(dy).$$

We may let $a \to 1$. Hence for all $t > c$ and x in B,

$$\|Q_n D^2 p_t f(x) - D^2 p_t f(x)\|_1 \leq \frac{\beta}{\sqrt{c}} \int_B \|Q_n y - y\| \, p_1(dy).$$

Finally,

$$\int_B \|Q_n y - y\| p_1(dy) = \int_{B_0} \| Q_n y - y\| \, p_1(dy)$$

$$\leq \int_{B_0} \| Q_n y - y\|_0 \, p_1(dy)$$

$$= \int_{B_0} \| Q_n y - y\|_0 \, \tilde{p}_1(dy)$$

$$\to 0 \text{ as } n \to \infty \quad \text{by Step 2.}$$

Hence we have

$$\lim_{n\to\infty} \|Q_n D^2 p_t f(x) - D^2 p_t f(x)\|_1 = 0 \quad \text{uniformly for } t > c \text{ and } x \in B.$$

<u>(c)</u>. $v(t, x) = p_t f(x) = \int_B f(x+y) p_t(dy)$

$$= \int_B f(x+\sqrt{t}y) p_1(dy).$$

Hence

$$|v(t,x)-v(s,y)| \leq \beta \int_B (\|x-y\| + |\sqrt{t}-\sqrt{s}| \|z\|) p_1(dz)$$

$$= \beta\{\|x-y\| + |\sqrt{t}-\sqrt{s}| \int_B \|z\| p_1(dy)\}.$$

<u>(d)</u>. <u>Step 1</u>:

$$p_{t+\varepsilon} f(x) = \int_B p_{\varepsilon^4(t+\varepsilon)} f(x+y) p_{(1-\varepsilon^4)(t+\varepsilon)}(dy)$$

$$= \int_B p_{\varepsilon^4(t+\varepsilon)} f(x+\sqrt{(1-\varepsilon^4)(t+\varepsilon)}y) p_1(dy).$$

Similarly,

$$p_t f(x) = \int_B p_{\varepsilon^4 t} f(x+\sqrt{(1-\varepsilon^4)t}y) p_1(dy).$$

Let

$$\theta(\varepsilon,x) = \int_B (p_{\varepsilon^4(t+\varepsilon)} f - p_{\varepsilon^4 t} f)(x+\sqrt{(1-\varepsilon^4)(t+\varepsilon)}y) p_1(dy)$$

and

$$\rho(\varepsilon,x) = \int_B [p_{\varepsilon^4 t} f(x+\sqrt{(1-\varepsilon^4)(t+\varepsilon)}y) - p_{\varepsilon^4 t} f(x+\sqrt{(1-\varepsilon^4)t}y)] p_1(dy).$$

Then

$p_{t+\varepsilon}f(x)-p_t f(x)= \theta(\varepsilon,x)+\rho(\varepsilon,x).$

Step 2: $(p_{\varepsilon^4(t+\varepsilon)}f-p_{\varepsilon^4 t}f)(z)= \int_B [f(z+\varepsilon^2\sqrt{t+\varepsilon}y)-f(z+\varepsilon^2\sqrt{t}y)]p_1(dy).$

Hence, for all $z \in B$,

$$|(p_{\varepsilon^4(t+\varepsilon)}f-p_{\varepsilon^4 t}f)(z)|\leq\beta\varepsilon^2(\sqrt{t+\varepsilon}-\sqrt{t}) \int_B ||y||p_1(dy).$$

Therefore

$$\lim_{\varepsilon\to 0} \frac{1}{\varepsilon}|\theta(\varepsilon,x)|= 0 \quad \text{uniformly in } x \in B.$$

Step 3: Recall that $Dp_s f(x) \in B^*$ for all $x \in B$. Hence $\rho(\varepsilon,x)$ can be written as

$$\rho(\varepsilon,x)= \int_0^1\int_B \lambda(t,\varepsilon)(Dp_{\varepsilon^4 t}f(x+\sqrt{(1-\varepsilon^4)t}y+s\lambda(t,\varepsilon),y)p_1(dy)ds,$$

where $\lambda(t,\varepsilon)=\sqrt{(1-\varepsilon^4)(t+\varepsilon)}-\sqrt{(1-\varepsilon^4)t}.$

On the other hand, we may let $\tilde{T} \to I$ in Lemma 4.4 and apply the Lebesgue dominated convergence theorem to get

$$\text{trace } D^2p_t f(x)= \frac{1}{\sqrt{(1-\varepsilon^4)t}}\int_B(Dp_{\varepsilon^4 t}f(x+\sqrt{(1-\varepsilon^4)t}y),y)p_1(dy).$$

For each $t > 0$ fixed, it follows from (b) that trace $D^2p_t f(.)$ is uniformly continuous. Using this fact and the above two equations for ρ and trace $D^2p_t f(x)$, we can easily check that

$$\lim_{\varepsilon \to 0} \left|\frac{1}{\varepsilon}\rho(\varepsilon,x) - \frac{1}{2}\text{ trace } D^2 p_t f(x)\right| = 0 \text{ uniformly in } x \in B.$$

Therefore, from step 2 we have

$$\lim_{\varepsilon \to 0} \left|\frac{1}{\varepsilon}(p_{t+\varepsilon}f(x) - p_t f(x)) - \frac{1}{2}\text{ trace } D^2 p_t f(x)\right| = 0$$

uniformly in $x \in B$. This shows (d). #

Lemma 4.5. Let $b > 0$. Then for any n

$$p_t(\|y\| < b) = O(t^{-n/2}), \quad t \to \infty.$$

Proof. Let $\{e_1, \ldots, e_n\} \subset B^*$ be orthonormal in H. Let K be the span of $\{e_1, \ldots, e_n\}$. Let $\alpha = \sup\{|(y,e_j)|;\ \|y\| < b,\ j=1,\ldots,n\}$. Then $(\|y\| < b)$ is contained in the cylinder set

$$C = \{y \in B;\ |(y,e_j)| < \alpha,\ j=1,\ldots,n\}.$$

Hence

$$p_t(\|y\| < b) \le p_t(C)$$

$$= (2\pi t)^{-n/2} \int_A e^{-|x|^2/2t}\, dx,$$

where dx is the Lebesgue measure in K and $A = \{x \in K;\ |(x,e_j)| < \alpha,\ j=1,\ldots,n\}$. The conclusion follows immediately. #

Lemma 4.6. Let f be a bounded continuous function in B with bounded support. Then for each x in B

$$\int_0^\infty Dp_t f(x)dt \text{ and } \int_0^\infty D^2 p_t f(x)dt$$

converge in H and in $L(H)$, respectively.

Proof. From Chapter II Theorem 6.2 we have

$$\langle Dp_t f(x), h\rangle = \frac{1}{t}\int_B f(x+y)\langle y, h\rangle p_t(dy).$$

Hence

$$|\langle Dp_t f(x), h\rangle| \le \frac{1}{t}\left[\int_B f(x+y)^2 p_t(dy)\right]^{1/2}\left[\int_B \langle y,h\rangle^2 p_t(dy)\right]^{1/2}$$
$$= \frac{1}{t}\left[\int_B f(x+y)^2 p_t(dy)\right]^{1/2}\sqrt{t}|h|.$$

Therefore, we have

$$|Dp_t f(x)| \le \frac{1}{\sqrt{t}}\left[\int_B f(x+y)^2 p_t(dy)\right]^{1/2}.$$

On one hand, we have

$$|Dp_t f(x)| \le \frac{1}{\sqrt{t}}||f||_\infty,$$

which implies that

$$\int_0^1 |Dp_t f(x)|dt < \infty.$$

On the other hand, when x is fixed, $f(x+\cdot)$ vanishes outside $(\|y\| < b)$, say, since f has bounded support.

Then

$$|Dp_t f(x)| \leq \frac{1}{\sqrt{t}} [\int_{\|y\|<b} f(x+y)^2 p_t(dy)]^{1/2}$$

$$\leq \frac{1}{\sqrt{t}} \|f\|_\infty [p_t(\|y\| < b)]^{1/2}.$$

By Lemma 4.5

$$|Dp_t f(x)| = O(t^{-1/2 - n/4}) \quad \text{for any } n.$$

Thus we can choose $n > 2$, say, so that $t^{-1/2 - n/4}$ is integrable at ∞. Finally, it is easy to check that $Dp_t f(x)$ is continuous in t. Hence

$$\int_1^\infty |Dp_t f(x)| dt < \infty.$$

Hence the integral $\int_0^\infty Dp_t f(x) dt$ converges in H. The second conclusion can be shown in a similar way. #

Definition 4.4. The Green measure on B is the Borel measure defined by

$$G(A) = \int_0^\infty p_t(A) dt, \quad A \in \mathcal{B}(B).$$

The potential of a measurable function f in B is the function defined by

$$Gf(x) = \int_B f(x+y) G(dy)$$

when it exists.

Theorem 4.7. Let f be a bounded Lip-1 function in B with bounded support. Then $D^2Gf(x) \in \mathcal{L}_{(1)}(H)$ for all x in B and

$$\text{trace } D^2Gf(x) = -2f(x).$$

Proof. Clearly,

$$Gf(x) = \int_0^\infty p_t f(x)\,dt.$$

Using Lemma 4.6, we can check that Gf is twice H-differentiable with the second derivative given by

$$D^2Gf(x) = \int_0^\infty D^2 p_t f(x)\,dt \in \mathcal{L}(H).$$

From Theorem 4.6, $D^2 p_t f(x) \in \mathcal{L}_{(1)}(H)$. We show that the integral $\int_0^\infty ||D^2 p_t f(x)||_1\,dt$ converges, so that $D^2Gf(x) \in \mathcal{L}_{(1)}(H)$.

In the proof of Theorem 4.6 (a) we have the estimate

$$\|D^2 p_t f(x)\|_1 \le \beta t^{-1/2} \int_B \|y\|\, p_1(dy).$$

Hence

$$\int_0^1 \|D^2 p_t f(x)\|_1\,dt < \infty.$$

On the other hand, when x is fixed, $f(x+\cdot)$ vanishes, say, outside $\{y \in B;\ \|y\| < \frac{\alpha}{2}\}$ for some α since f has bounded support. Then

$$\langle Dp_s f(x), h\rangle = \lim_{\varepsilon\to 0} \frac{1}{\varepsilon}[p_s f(x+\varepsilon h) - p_s f(x)]$$

$$= \lim_{\varepsilon\to 0} \int_B \frac{1}{\varepsilon}[f(x+\varepsilon h+y) - f(x+y)] p_s(dy)$$

$$= \lim_{\varepsilon\to 0} \int_{||y||<\alpha} \frac{1}{\varepsilon}[f(x+\varepsilon h+y) - f(x+y)] p_s(dy).$$

Hence, using the Lipschitzian condition $|f(x)-f(y)| \leq \beta||x-y||$, we have

$$\|Dp_s f(x)\|_* \leq \beta p_s(||y|| < \alpha), \text{ for all } x \in B.$$

Use this estimate and Lemma 4.4, and then carry out the same computation in the proof of Theorem 4.6 (a), it is easy to obtain the estimate

$$||D^2 p_t f(x)||_1 \leq \beta (at)^{-1/2} p_{bt}(||y|| < \alpha) \int_B ||y|| p_1(dy),$$

where $a > 0$, $b > 0$ and $a+b = 1$. Choosing $a=b=\frac{1}{2}$, we have

$$||D^2 p_t f(x)||_1 \leq \beta\sqrt{2}\, t^{-1/2} p_{\frac{t}{2}}(||y|| < \alpha) \int_B ||y|| p_1(dy).$$

It follows from Lemma 4.5 that for any n

$$||D^2 p_t f(x)||_1 = O(t^{-1/2-n/2}), \ t \to \infty.$$

That we can choose n large so that $t^{-1/2-n/2}$ is integrable at $t = \infty$. Moreover, $D^2 p_{(.)} f(.)$ is uniformly continuous from $[1,\infty) \times B$ into $L_{(1)}(H)$ by Theorem 4.6 (b). Therefore,

$$\int_1^\infty \|D^2 p_t f(x)\|_1 dt < \infty.$$

Hence

$$\int_0^\infty \|D^2 p_t f(x)\|_1 dt < \infty, \text{ and so } D^2Gf(x) \in L_{(1)}(H)$$

for all x in B.

Finally, we have

$$\text{trace } D^2Gf(x) = \int_0^\infty \text{trace } D^2 p_t f(x)\, dt$$

$$= \lim_{\varepsilon \downarrow 0, R \uparrow \infty} \int_\varepsilon^R \text{trace } D^2 p_t f(x)\, dt$$

$$= \lim_{\varepsilon \downarrow 0, R \uparrow \infty} \int_\varepsilon^R 2\partial p_t f(x)/\partial t\, dt \quad \text{(by Theorem 4.6 (d))}$$

$$= 2 \lim_{\varepsilon \downarrow 0, R \uparrow \infty} [p_R f(x) - p_\varepsilon f(x)]$$

$$= -2f(x). \qquad \#$$

Piech [37] has generalized Theorem 4.6 to the parabolic equation

$$\partial u(t,x)/\partial t = \text{trace } A(x) D^2 u(t,x),$$

where $A(x) = I - C(x)$ satisfies the following conditions:

(i) C is a map from B into the symmetric trace class operators of H,

(ii) there exists $\varepsilon > 0$ such that $A(x) \geq \varepsilon I$ for all x in B,

(iii) there exists a symmetric Hilbert-Schmidt operator E of H and a family of operators $C_0(x) \in L(H)$ such that $C(x) = EC_0(x)E$ and $\|C_0(x)\| \leq 1$ for all x in B.

(iv) C_0 is twice Fréchet differentiable and $C_0''(x)$ is a Lip-1 function,

(v) $\|C_0'(x)\|$ and $\|C_0''(x)\|$ are uniformly bounded for all x in B.

Theorem 4.8. Under the above conditions (i)–(v), there exists a family of Borel meausres $\{q_t(x, dy);\ t > 0,\ x \in B\}$ such that if f is a bounded Lip-1 function in B then

(a) $q_t f(x) \equiv \int_B f(y) q_t(x, dy)$ is differentiable in t, twice H-differentiable in x and $D^2 q_t f(x) \in L_{(1)}(H)$ for all x in B,

(b) $\partial q_t f(x)/_{\partial t} = \text{trace } A(x) D^2 q_t f(x)$,

(c) $\lim_{t \downarrow 0} q_t f(x) = f(x)$ uniformly in x.

§ 5. Stochastic integral.

Let (i, H, B) be a fixed abstract Wiener space. By Chapter I Corollary 4.2 there exist an abstract Wiener space (j, H, B_0) and an increasing sequence $\{Q_n;\ n=1,2,\ldots\ldots\}$ of finite dimensional orthogonal projections of H such that (i) $\|\cdot\|$ is weaker than the B_0-norm $\|\cdot\|_0$, (ii) Q_n converges strongly to I in H, (iii) each Q_n extends to a projection, still denoted by Q_n, of B_0 such that $\|Q_n\|_{B_0, B_0} \leq 1$ and (iv) $\|Q_n x - x\|_0 \to 0$ as $n \to \infty$ for each $x \in B_0$. We then have the following situation $B^* \subset B_0^* \subset H \subset B_0 \subset B$. Throughout this section $\{Q_n\}$ will denote the above sequence. $\{p_t\}$ and $\{\tilde{p}_t\}$ will denote the Wiener measures of B and

B_0, respectively.

Let K be a real Hilbert space. The same argument in the proof of Chapter I Corollary 4.4 can be used to prove the following lemma.

Lemma 5.1. If $A \in L(B_0, K)$ and $\tilde{A}$ is the restriction of A to H, then $\tilde{A} \in L_{(2)}(H, K)$ (the Hilbert-Schmidt operators from H into K). In fact

$$||\tilde{A}||_2 \leq \left(\int_{B_0} ||x||_0^2 \tilde{p}_1(dx)\right)^{1/2} ||A||_{B_0,K}, \quad A \in L(B_0, K),$$

where $||.||_2$ denotes the Hilbert-Schmidt norm in $L_{(2)}(H, K)$.

Lemma 5.2. $L(B_0, K)$ is dense in $L_{(2)}(H, K)$.

Proof. If $A \in L_{(2)}(H, K)$, define $A_n = AQ_n$. Then $A_n \in L(B_0, K)$ and $||A_n - A||_2 \to 0$ as $n \to \infty$. #

Lemma 5.3. If $A \in L(B_0, K)$ then $\tilde{A}^*(K) \subset B_0^*$ and

$$\langle Ax, y\rangle_K = (x, \tilde{A}^* y) \text{ for all } x \in B_0 \text{ and } y \in K.$$

Let W(t) be a standard Wiener process in (i, H, B). Note that Prob $\{\omega;\ W(t,\omega) \in B_0$ for all $t \geq 0\} = 1$. Let M_t be the complete σ-field generated by $\{W(s);\ 0 \leq s \leq t\}$.

Definition 5.1. Let G be a Banach space. A stochastic process $\zeta(t)$ with state space G is called nonanticipating if ζ is (t,ω)-

jointly measurable and for each $t \geq 0$, $\zeta(t)$ is M_t-measurable. $L[G]$ will denote the space of nonanticipating processes $\zeta(t)$ with state space G such that $E\int_0^s \|\zeta(t)\|_G^2 dt < \infty$ for each $0 < s < \infty$.

Lemma 5.4. If $\zeta \in L[L(B_0, K)]$ then we have

(a) for $s < t$, $E|\zeta(s)(W(t)-W(s))|_K^2 = (t-s)E[\|\tilde{\zeta}(s)\|_2^2]$, and

(b) for $s < t < u < v$, $E<\zeta(s)(W(t)-W(s)), \zeta(u)(W(v)-W(u))>_K = 0$.

Proof. We prove (a) only. (b) can be proved similarly. Let

$$f = |\zeta(s)(W(t)-W(s))|_K^2$$

and

$$f_n = |\zeta(s)Q_n(W(t)-W(s))|_K^2, \quad n=1,2,\ldots\ldots$$

Then f_n converges to f almost surely as $n \to \infty$. Moreover

$$f_n \leq \|\zeta(s)Q_n\|_{B_0,K}^2 \|W(t)-W(s)\|_0^2$$

$$\leq \|\zeta(s)\|_{B_0,K}^2 \|W(t)-W(s))\|_0^2 .$$

But since ζ is nonanticipating, we have

$$E[\|\zeta(s)\|_{B_0,K}^2 \|W(t)-W(s))\|_0^2]$$

$$= E(\|\zeta(s)\|_{B_0,K}^2) E(\|W(t)-W(s))\|_0^2)$$

$$=(t-s)E(\|\zeta(s)\|^2_{B_0,K})\int_{B_0}\|x\|_0^2\tilde{P}_1(dx).$$

By the Lebesgue dominated convergence theorem we then have

$$E(f)=\lim_{n\to\infty}E(f_n).$$

Let $\{d_k;\ k=1,2,\cdots\}$ be an orthonormal basis of K. The monotone convergence theorem yields

$$E(f_n)=\lim_{k\to\infty}\sum_{j=1}^{k}E\langle\zeta(s)Q_n(W(t)-W(s)),\ d_j\rangle_K^2$$

$$=\lim_{k\to\infty}\sum_{j=1}^{k}E(Q_n(W(t)-W(s)),\ \tilde{\zeta}(s)^*d_j)^2, \quad \text{(by Lemma 5.3).}$$

Now, let $\{e_n;\ n=1,2,\cdots\}$ be an orthonormal basis of H such that $\{e_1,\cdots\cdots,e_{\ell(n)}\}$ is a basis for $Q_n(H)$. Then

$$(Q_n(W(t)-W(s)),\ \tilde{\zeta}(s)^*d_j)^2$$

$$=\sum_{\alpha,\beta=1}^{\ell(n)}(W(t)-W(s),e_\alpha)(W(t)-W(s),e_\beta)(e_\alpha,\tilde{\zeta}(s)^*d_j)(e_\beta,\ \tilde{\zeta}(s)^*d_j).$$

Using the assumption that ζ is nonanticipating and $E(W(t)-W(s),e_\alpha)(W(t)-W(s),e_\beta)=\delta_{\alpha\beta}(t-s)$, we obtain that

$$E(Q_n(W(t)-W(s)),\tilde{\zeta}(s)^*d_j)^2=(t-s)E\sum_{\alpha=1}^{\ell(n)}(e_\alpha,\tilde{\zeta}(s)^*d_j)^2$$

$$=(t-s)E\sum_{\alpha=1}^{\ell(n)}\langle\tilde{\zeta}(s)e_\alpha,d_j\rangle_K^2.$$

Therefore.

$$E(f_n)=\lim_{k\to\infty}\sum_{j=1}^{k}(t-s)E\sum_{\alpha=1}^{\ell(n)}\langle\tilde{\zeta}(s)e_\alpha, d_j\rangle_K^2$$

$$=(t-s)E\sum_{\alpha=1}^{\ell(n)}|\tilde{\zeta}(s)e_\alpha|_K^2.$$

And so

$$E(f)=\lim_{n\to\infty}E(f_n)$$

$$=(t-s)E\sum_{\alpha=1}^{\infty}|\tilde{\zeta}(s)e_\alpha|_K^2$$

$$=(t-s)E[\|\tilde{\zeta}(s)\|_2^2].$$ #

Definition 5.2. $\zeta\in L[G]$ is called simple if there is a finite set of points $\{0=t_0<t_1<\cdots<t_n\}$ such that $\zeta(t)=\zeta(t_j)$ for $t_j\le t<t_{j+1}$, $j=0,1,2,\cdots\cdots n-1$, and $\zeta(t)=\zeta(t_n)$ when $t\ge t_n$.

Lemma 5.5. If $\zeta\in L[\mathcal{L}_{(2)}(H,K)]$ then there exists a sequence $\{\zeta_n\}$ of simple processes in $L[\mathcal{L}(B_0,K]$ such that $E\int_0^s\|\zeta_n(t)-\zeta(t)\|_2^2\,dt\to 0$ as $n\to\infty$, for each $0<s<\infty$.

Let ζ be a simple process in $L[\mathcal{L}(B_0,K)]$ with jumps at $0<t_1<\cdots\cdots<t_n$. When $t_j\le t<t_{j+1}$, $0\le j\le n$, define

$$J_\zeta(t)=\sum_{i=0}^{j-1}\zeta(t_i)(W(t_{i+1})-W(t_i))+\zeta(t_j)(W(t)-W(t_j)).$$

Here $t_0=0$ and $t_{n+1}=\infty$ by convention.

Obviously, J_ζ is a martingale, has continuous sample paths and $E\,J_\zeta(t)=0$. Moreover, by Lemma 5.4

$$E|J_\zeta(t)|_K^2 = E\int_0^t \|\tilde{\zeta}(s)\|_2^2\,ds.$$

Suppose that $\zeta \in L[\mathcal{L}_{(2)}(H,K)]$. Let $\{\zeta_n\}$ be a sequence given in Lemma 5.5. For each $0 < s < \infty$, J_{ζ_n} converges to Θ_s in $L^2([0,s]\times\Omega)$. Obviously, if $s < v$ then the restriction of Θ_v to $[0,s]\times\Omega$ equals to Θ_s almost everywhere. Therefore, we can define a process Θ such that for almost all $t \in [0,\infty)$

$$\lim_{n\to\infty} E|J_{\zeta_n}(t) - \Theta(t)|_K^2 = 0.$$

Θ has an equivalent version J_ζ, which has the following properties.

Theorem 5.1. There exists a linear operator J from $L[\mathcal{L}_{(2)}(H,K)]$ into $L[K]$, denoted by $J_\zeta(t)=\int_0^t \zeta(s)dW(s)$, such that

(a) J_ζ has continuous sample paths,

(b) J_ζ is a martingale,

(c) $\mathrm{Prob}\{\sup_{0\le t\le s} |J_\zeta(t)|_K > \delta\} \le \delta^{-2} E|J_\zeta(s)|_K^2$,

(d) $EJ_\zeta(t)=0$ and $E|J_\zeta(t)|_K^2 = E\int_0^t \|\zeta(s)\|_2^2\,ds$.

Remark. When $K = \mathbb{R}$, we have $\mathcal{L}_{(2)}(H,\mathbb{R}) = H$ by convention. When $\zeta \in L[H]$, $J_\zeta(t)$ will be instead denoted by $\int_0^t \langle\zeta(s), dW(s)\rangle$.

Lemma 5.6. Let $\zeta \in L(\mathcal{L}(B_0, B_0^*))$. Then

(a) for $s < t$, $E\{(\zeta(s)(W(t)-W(s)), W(t)-W(s))^2\} = (t-s)^2 E\{2\|\tilde{\zeta}(s)\|_2^2$

$+(\mathrm{trace}\ \tilde{\zeta}(s))^2\}$,

(b) for $s < t \le u < v$, $E\{(\zeta(s)(W(t)-W(s)),W(t)-W(s))(\zeta(u)(W(v)-W(u)),W(v)-W(u))\}=(v-u)E\{\text{trace } \tilde{\zeta}(u)(\zeta(s)(W(t)-W(s)),W(t)-W(s))\}$.

Lemma 5.7. Let $\zeta \in L[L(B_0,B_0^*)]$ have continuous sample paths and $||\zeta(t,\omega)||_{B_0,B_0^*} \le M < \infty$ for all t and a.e. ω. Let $\pi_n = \{0=t_0 < t_1 < \cdots < t_n = t\}$ be a partition of $[0,t]$. Then

$$\sum_{j=1}^{n-1} (\zeta(t_j)(W(t_{j+1})- W(t_j)),\ W(t_{j+1})- W(t_j))$$

converges in $L^2(\Omega)$ to $\int_0^t \text{trace } \tilde{\zeta}(s)ds$ as mesh(π_n) tends to 0.

Let f be a function in an open subset V of B. The differentiability of f in the directions of B_0 (B_0-differentiability) can be defined in the same way as H-differentiability. If f is twice B_0-differentiable then $Df \in B_0^* \subset H$ and $D^2f \in L(B_0,B_0^*) \subset L_{(1)}(H)$.

Lemma 5.8. Let f be a twice B_0-differentiable function in B then

$$f(W(t))= f(0)+\int_0^t \langle Df(W(s)),dW(s)\rangle+ \frac{1}{2}\int_0^t \text{trace } D^2f(W(s))ds.$$

Proof. We have

$$f(x)-f(y)=(Df(y),x-y)+\frac{1}{2}(D^2f(y)(x-y),x-y)+o(||x-y||_0^2),$$

where x and y are in B_0. We may assume that $||D^2f||_{B_0,B_0^*}$ is bounded.

Let $\pi_n = \{0 = t_0 < t_1 < \cdots < t_n = t\}$ be a partition of $[0,t]$, then

$$f(W(t)) - f(0) = \sum_{j=0}^{n-1} f(W(t_{j+1})) - f(W(t_j))$$

$$= \sum_{j=1}^{n-1} (Df(W(t_j)), W(t_{j+1}) - W(t_j))$$

$$+ \frac{1}{2} \sum_{j=0}^{n-1} (D^2 f(W(t_j))(W(t_{j+1}) - W(t_j)), W(t_{j+1}) - W(t_j))$$

$$+ \sum_{j=0}^{n-1} o(||W(t_{j+1}) - W(t_j)||_0^2).$$

Using Theorem 5.1 (d), we can compute

$$E\left|\sum_{j=0}^{n-1} (Df(W(t_j)), W(t_{j+1}) - W(t_j)) - \int_0^t \langle Df(W(s)), dW(s)\rangle\right|^2$$

$$= \sum_{j=0}^{n-1} \int_{t_j}^{t_{j+1}} E|Df(W(t_j)) - Df(W(s))|^2 ds$$

$$\leq \text{constant} \sum_{j=0}^{n-1} \int_{t_j}^{t_{j+1}} E\| W(t_j) - W(s) \|_0^2 \, ds$$

$$= \text{constant} \sum_{j=0}^{n-1} \int_{t_j}^{t_{j+1}} (s - t_j) ds$$

$$\leq \text{constant} \sum_{j=0}^{n-1} (t_{j+1} - t_j)^2$$

$$\leq \text{constant } (\text{mesh } \pi_n) t \to 0 \quad \text{as mesh } \pi_n \to 0.$$

The above constant denotes any constant depending only upon $\sup_{x\in B}\|D^2f(x)\|_{B_0,B_0^*}$, $\int_{B_0}\|y\|^2\tilde{p}_1(dy)$ and a number β such that $\|y\|_0 \le \beta\,|y|$ for all $y \in H$ (so that $|z| \le \beta\,\|z\|_{0*}$ for all $z \in B_0^*$).

By Lemma 5.7

$$E\Big|\sum_{j=0}^{n-1}(D^2f(W(t_j))(W(t_{j+1})-W(t_j)),W(t_{j+1})-W(t_j)) - \int_0^t \text{trace}\, D^2f(W(s))ds\Big|^2$$

tends to zero as mesh $(\pi_n) \to 0$.

Moreover,

$$E\Big|\sum_{j=0}^{n-1}\{\|W(t_{j+1})-W(t_j)\|_0^2 - \int_{B_0}\|y\|^2\tilde{p}_1(dy)(t_{j+1}-t_j)\}\Big|^2$$

$$=[\int_{B_0}\|y\|^4\tilde{p}_1(dy)-(\int_{B_0}\|y\|^2\,\tilde{p}_1(dy))^2]\sum_{j=0}^{n-1}(t_{j+1}-t_j)^2$$

$$\longrightarrow 0 \quad \text{as mesh } (\pi_n) \to 0.$$

Hence

$$\sum_{j=0}^{n-1} o(\|W(t_{j+1})-W(t_j)\|_0^2) \to 0 \text{ in } L^2(\Omega) \text{ as mesh } \pi_n \to 0.$$

Finally, we can choose a subsequence of $\{\pi_n\}$ so that the above $L^2(\Omega)$-convergence can be replaced by a.s.-convergence with respect to this subsequence. Therefore, we have proved that the formula is true for each t fixed. Observe that integrals in this formula have continuous versions. Hence the formula holds for all $t > 0$. #

Theorem 5.2. (Ito's lemma). Let $f(t,x)$ be a real-valued continuous function on $[0, \infty)\times B$. Assume that

(a) for each $x \in B$, $f(\cdot, x)$ is continuously differentiable and $\partial f/\partial t$ is continuous on $[0, \infty)\times B$,

(b) for each $t \geq 0$, $f(t, \cdot)$ is twice H-differentiable such that Df and D^2f are continuous from $[0, \infty)\times B$ into H and $L(H)$, respectively. If

$$X(t)= x + \int_0^t \zeta(s)dW(s)+ \int_0^t \sigma(s)ds,$$

where $x \in B$, $\zeta \in L[L_{(2)}(H)]$ and $\sigma \in L[H]$. Then

$$f(t,X(t))=f(0,x)+ \int_0^t \langle \zeta(s)^* Df(s,X(s)),dW(s)\rangle$$
$$+\int_0^t \{\frac{\partial f}{\partial s}(s,X(s))+\langle Df(s,X(s)),\sigma(s)\rangle$$
$$+ \frac{1}{2} \operatorname{trace}[\zeta^*(s)D^2f(s,X(s))\zeta(s)]\}ds.$$

Proof. The same argument in the proof of Lemma 5.8 can be used to show that the above conclusion is valid for $f(t,x)$, where f satisfies (a) and (b) with H replaced by B_0.

Now, suppose f satisfies (a) and (b). For each n, consider the function $f_n(t,x)=f(t,Q_n x)$. Then $\partial f_n/\partial t\,(t,x)=\partial f/\partial t\,(t,Q_n x)$, $Df_n(t,x)=Q_n f(t,Q_n x)$ and $D^2f_n(t,x)=Q_n D^2f(t,Q_n x)Q_n$. Hence by the preceeding paragraph we can apply f_n to the theorem. Letting $n \to \infty$, we obtain the conclusion for f. #

If $\xi(s) = I + \zeta(s)$, where I is the identity map of B and $\zeta \in L[\mathcal{L}_{(2)}(H)]$, we define

$$\int_0^t \xi(s)dW(s) = W(t) + \int_0^t \zeta(s)dW(s).$$

<u>Theorem 5.3.</u> (Ito's lemma). Let $f(t,x)$ be a real-valued continuous function on $[0,\infty)\times B$ satisfying condition (a) in Theorem 5.2 and the following condition

(b)' for each $t \geq 0$, $f(t,\cdot)$ is twice H-differentiable such that $D^2f(t,x) \in \mathcal{L}_{(1)}(H)$ for each x and $Df : [0,\infty)\times B \to H$ and $D^2f : [0,\infty)\times B \to \mathcal{L}_{(1)}(H)$ are continuous.

If $X(t) = x + \int_0^t \xi(s)dW(s) + \int_0^t \sigma(s)ds$, where $\xi = I+\zeta$, $\zeta \in L[\mathcal{L}_{(2)}(H)]$ and $\sigma \in L[H]$. Then

$$f(t,X(t)) = f(0,x) + \int_0^t \langle \tilde{\xi}(s)^* Df(s,X(s)), dW(s)\rangle$$

$$+ \int_0^t \{\frac{\partial f}{\partial s}(s,X(s)) + \langle Df(s,X(s)), \sigma(s)\rangle + \frac{1}{2}\,\text{trace}\,[\tilde{\xi}(s)^* D^2 f(s,X(s))\tilde{\xi}(s)]\}ds.$$

<u>Corollary 5.1.</u> Let f be a twice continuously H-differentiable function in B such that $D^2f(x) \in \mathcal{L}_{(1)}(H)$ for all $x \in B$ and D^2f is continuous from B into $\mathcal{L}_{(1)}(H)$. Let τ be a stopping time

(w.r.t. $W(t)$) such that $\tau < \infty$ a.s. and $\int_1^\infty E[1_{t\leq\tau}|Df(x_0+W(t))|^2]dt < \infty$. Then

$$E[f(x_0+W(\tau))]=f(x_0)+E[\int_0^\tau \frac{1}{2}\text{ trace } D^2f(x_0+W(t))dt].$$

Corollary 5.2. Let $A \in L(B,B^*)$ and $u(x)=\frac{1}{2}(Ax,x)$. Then $\Delta u(x)=$ trace $(A|_H)$ for all x in B, where Δ is the generalized Laplacian.

Definition 5.3. Let K be a Hilbert space. Let G be the completion of K with respect to a weaker norm in K. The couple (K, G) is called a conditional Banach space.

Obviously, an abstract Wiener space is a conditional Banach space. But the converse is not true, e.g., $(L^2[0,1], L^1[0,1])$ is a conditional Banach space, but it is not an abstract Wiener space.

Definition 5.4. Let K_1 and K_2 be two Hilbert spaces. A continuous bilinear map S from $K_1 \times K_1$ into K_2 is called trace-class-type if (i) for each $x \in K_2$, $S_x \in L_{(1)}(K_1)$, where $\langle S_x y,z\rangle=\langle S(y,z),x\rangle$, and (ii) the linear functional $x \mapsto$ trace S_x is continuous.

It follows from the definition that there exists a unique vector, denoted by TRACE S, in K_2 such that $\langle$TRACE S, $x\rangle=$ trace S_x for all x in K_2.

Lemma 5.9. If $T \in L_{(2)}(H,K_1)$ and S is a continuous bilinear map from $K_1 \times K_1$ into K_2. Then the bilinear map $S\circ[T \times T]$ from $H\times H$ into K_2 is trace-class-type.

<u>Theorem 5.4</u>. (Ito's lemma). Let (K_1, G_1) and (K_2, G_2) be two conditional Banach spaces. Let ρ be a function from $[0,\infty)$ $\times G_1$ into G_2 satisfying the following conditions:

(I) for each x in G_1, $\rho(\cdot, x)$ is continuously differentiable and $\partial f/\partial t$ is continuous from $[0,\infty)\times G_1$ into G_2,

(II) for each $t \geq 0$, $\rho(t, \cdot): G_1 \to G_2$ is twice Fréchet differentiable such that ρ' and ρ'' are (t,x)-jointly continuous, $\rho'(t,x)(K_1) \subset K_2$ and $\rho''(t,x)(K_1 \times K_1) \subset K_2$.

If $X(t) = x + \int_0^t \zeta(s)dW(s) + \int_0^t \sigma(s)ds$,

where $x \in G_1$, $\zeta \in L[\mathcal{L}_{(2)}(H,K_1)]$ and $\sigma \in L[G_1]$. Then

$$\rho(t,X(t)) = \rho(0,x) + \int_0^t [\rho'(s,X(s)) \circ \zeta(s)]dW(s)$$

$$+ \int_0^t \{\frac{\partial \rho}{\partial s}(s,X(s)) + \rho'(s,X(s))(\sigma(s)) + \frac{1}{2} \text{TRACE}$$

$$(\rho''(s,X(s)) \circ [\zeta(s) \times \zeta(s)])\}ds.$$

If $\xi(s) = J + \zeta(s)$, where $J \in \mathcal{L}(B_0, G_1)$ and $\zeta \in L[\mathcal{L}_{(2)}(H,K_1)]$ we define

$$\int_0^t \xi(s)dW(s) = JW(t) + \int_0^t \zeta(s)dW(s).$$

<u>Theorem 5.5</u> (Ito's lemma). Let (K_1,G_1) and (K_2,G_2) be two conditional Banach spaces. Assume that G_1 has a Schauder basis. Let ρ be a function from $[0, \infty)\times G_1$ into G_2 satisfying conditio

(I) in Theorem 5.4 and the following condition:

(II)' for each $t \geq 0$, $\rho(t, \cdot)$ is twice K_1-differentiable such that $D\rho(t,x) \in \mathcal{L}(K_1, K_2)$ and $D^2\rho(t, x)$ is a trace-class-type bilinear map from $K_1 \times K_1$ into K_2. Also, $D\rho:[0,\infty) \times G_1 \to \mathcal{L}(K_1, K_2)$ and TRACE $D^2\rho :[0,\infty)\times G_1 \to K_2$ are continuous.

If $X(t)= x + \int_0^t \xi(s)dW(s)+ \int_0^t \sigma(s)ds$,

where $\xi = J + \zeta$, $J \in \mathcal{L}(B_0, G_1)$ and $\zeta \in L[\mathcal{L}_{(2)}(H, K_1)]$, and $\sigma \in L[K_1]$. Suppose $D\rho(t, x)(J(B_0)) \subset K_2$ for all t and x. Then

$$\rho(t,X(t))=\rho(0,x)+\int_0^t [D\rho(s,X(s))\circ\xi(s)]dW(s)$$

$$+\int_0^t \{\frac{\partial\rho}{\partial s}(s,X(s))+D\rho(s,X(s))(\sigma(s))+\frac{1}{2} \text{TRACE}$$

$$(D^2\rho(s,X(s))\circ[\xi(s)\times\xi(s)])\}ds.$$

Lemma 5.10. Let Φ be a map from a complete metric space $\mathfrak{A}$ into itself. Suppose there exists a natural number N such that Φ^m is a contraction map whenever $m \geq N$. Then Φ has a unique fixed point.

Let $\xi(t,x)= J + \zeta(t, x)$, where $J \in \mathcal{L}(B_0)$ and ζ is a map from $[a, \infty)\times B$ into $\mathcal{L}_{(2)}(H)$, $a \geq 0$. Let σ be a map from $[a,\infty) \times B$ into B. Consider the stochastic integral equation

$$X(t)= \nu + \int_a^t \xi(s,X(s))dW(s)+ \int_a^t \sigma(s,X(s))ds.$$

Theorem 5.6. Assume that ζ and σ satisfy the following conditions:

(a) For each x in B, $\zeta(\cdot, x)$ and $\sigma(\cdot, x)$ are continuous functions from $[a, \infty)$ into $L_{(2)}(H)$ and B, respectively.

(b) There exists a constant K such that for all $t \geq a$ and $x,y \in B$,

$$\|\zeta(t,x)-\zeta(t,y)\|_2 \leq K\|x-y\|$$

$$\|\sigma(t,x)-\sigma(t,y)\| \leq K\|x-y\|$$

$$\|\zeta(t,x)\|_2 \leq K(1+\|x\|)$$

$$\|\sigma(t,x)\| \leq K(1+\|x\|).$$

Suppose ν is M_a-measurable and $E\|\nu\|^2 < \infty$. Let $\xi(t,x) = J + \zeta(t,x)$, $J \in L(B_0)$. Then the stochastic integral equation

$$X(t) = \nu + \int_a^t \xi(s,X(s))dW(s) + \int_a^t \sigma(s,X(s))ds$$

has a unique nonanticipating continuous solution. Moreover, the solution is a Markov process.

Proof. We need only to consider the case $a \leq t \leq T < \infty$. Let $\mathfrak{A}$ be the Banach space of nonanticipating continuous processes X(t) with state space B such that

$$|||X||| = \{\sup_{a \leq t \leq T} E\|X(t)\|^2\}^{1/2} < \infty.$$

Define a map Φ from $\mathfrak{A}$ into itself by

$$\Phi(X)(t) = \nu + \int_a^t \xi(s,X(s))dW(s) + \int_a^t \sigma(s,X(s))ds.$$

Let c be a constant such that $\|x\| \leq c|x|$ for x in H.

Then

$$\|\Phi(X)(t) - \Phi(Y)(t)\|^2 \leq 2c^2 \left|\int_a^t [\zeta(s,X(s)) - \zeta(s,Y(s))]dW(s)\right|^2$$

$$+ 2(t-a)\int_a^t \|\sigma(s,X(s)) - \sigma(s,Y(s))\|^2 ds.$$

Hence

$$E\|\Phi(X)(t) - \Phi(Y)(t)\|^2 \leq 2c^2 \int_a^t E\|\zeta(s,X(s)) - \zeta(s,Y(s))\|^2 ds$$

$$+ 2(T-a)\int_a^t E\|\sigma(s,X(s)) - \sigma(s,Y(s))\|^2 ds$$

$$\leq 2K^2(c^2+T-a)\int_a^t E\|X(s) - Y(s)\|^2 ds.$$

Let $\alpha = 2K^2(c^2+T-a)$. Then

$$E\|\Phi(X)(t) - \Phi(Y)(t)\|^2 \leq \alpha\int_a^t E\|X(s) - Y(s)\|^2 ds.$$

It is easy to see that for m > 1 we have

$$E\|\Phi^m(X)(t) - \Phi^m(Y)(t)\|^2 \leq \frac{\alpha(t-a)^m}{m!} |||X - Y|||^2 .$$

Hence

$$|||\ \Phi^m(X)-\Phi^m(Y)\ |||\ \leq\ [\frac{\alpha(T-a)^m}{m!}]^{1/2}\ \ |||\ X - Y\ |||.$$

Therefore Φ^m is a contraction map for large m. Then apply Lemma 5.10 to get a solution. Uniqueness of solution is obvious.

#

Suppose C is a symmetric compact operator of H such that I - C is positive definite. Let A = I - C. We define $\sqrt{A}$ as follows. C can be represented by

$$Cx = \sum_{n=1}^{\infty} \lambda_n <x, e_n> e_n,$$

where $\{e_n\}$ is an orthonormal basis of H. Then

$$Ax = \sum_{n=1}^{\infty} (1 - \lambda_n) <x, e_n> e_n.$$

Since A is positive definite, $1 - \lambda_n \geq 0$. Define

$$\sqrt{A}\ x = \sum_{n=1}^{\infty} \sqrt{1 - \lambda_n}\ <x, e_n>\ e_n.$$

The following lemma is easy to prove.

Lemma 5.11. Suppose C is symmetric.

(a) If $C \in \mathcal{K}(H)$ then $\sqrt{A} - I \in \mathcal{K}(H)$,

(b) If $C \in \mathcal{L}_{(2)}(H)$ then $\sqrt{A} - I \in \mathcal{L}_{(2)}(H)$,

(c) If $C \in \mathcal{L}_{(1)}(H)$ then $\sqrt{A} - I \in \mathcal{L}_{(1)}(H)$.

Consider now the parabolic equation

$$\partial v(t, x)/_{\partial t} = \text{trace } A(x)D^2v(t, x),$$

where $A(x) = I - C(x)$, and C satisfies the conditions in Theorem 4.8. Since $C(x)$ is a symmetric trace class operator of H, $\sqrt{A(x)} - I \in \mathcal{L}_{(1)}(H)$, by the above lemma. Moreover, it is easy to see that there exists a constant K such that

$$||\sqrt{A(x)} - \sqrt{A(y)}||_2 \leq K||x - y||$$

$$||\sqrt{A(x)} - I||_2 \leq K(1 + ||x||).$$

Therefore, by Theorem 5.6 the stochastic integral equation

$$X(t) = x + \int_0^t \sqrt{A(X(s))}\, dW(s)$$

has a unique solution $X(t)$ for each x in B.

<u>Theorem 5.7</u>. Let $\{q_t(x,dy);\ t > 0,\ x \in B\}$ be the fundamental solution of the above parabolic equation. Let $X(t)$ be the solution of the above stochastic integral equation. Then

$$q_t(x, dy) = \text{Prob}\{X(2t) \in dy | X(0) = x\}.$$

Proof. It is sufficient to show that for $T > 0$ we have

$$\int_B f(y) q_{T/2}(x, dy) = E_x f(X(T))$$

for any bounded Lip-1 function f. Let $T > 0$ be fixed. Define

$$F(t,x) = \int_B f(y) q_{(T-t)/2}(x,dy), \quad 0 \le t < T.$$

By Theorem 4.8 the function

$$g(t,x) \equiv \partial F(t,x)/\partial t + \frac{1}{2} \text{ trace } A(x) D^2 F(t,x)$$

is identically zero for $0 \le t < T$ and $x \in B$. On the other hand, upon applying Theorem 5.3 (Ito's Lemma) to $F(t,x)$, we have

$$F(t,X(t)) = F(0,x) + \int_0^t g(s,X(s))ds$$

$$+ \int_0^t \langle \sqrt{A(X(s))}\, DF(s,X(s)), dW(s) \rangle$$

$$= \int_B f(y) q_{T/2}(x,dy) + \int_0^t \langle \sqrt{A(X(s))}\, DF(s,X(s)), dW(s) \rangle .$$

Hence

$$E_x[F(t,X(t))] = \int_B f(y) q_{T/2}(x,dy).$$

Letting $t \uparrow T$ and using the Lebesgue dominated convergence theorem, we have

$$E_x[f(X(T))] = \int_B f(y) q_{T/2}(x, dy). \qquad \#$$

Let C_0 be the Banach space of bounded continuous functions in B vanishing at infinity. The norm of C_0 is the supremum norm. We assume that the function $f(x) = \|x\|^2$ is twice H-differentiable such that $Df: B \to H$ and $D^2f: B \to \mathcal{L}_{(1)}(H)$ are continuous and $\sup_{x \in B} \|D^2 f(x)\|_1 < \infty$. Consider the stochastic integral equation

$$X(t) = x + \int_0^t \xi(X(s))\,dW(s) + \int_0^t \sigma(X(s))\,ds,$$

where ξ and σ are independent of t and satisfy the conditions in Theorem 5.6. X(t) generates a semi-group $\{P_t;\ t \geq 0\}$, $P_t f(x) = E_x[f(X(t))]$, $f \in C_0$.

<u>Theorem 5.8.</u> The operators $\{P_t;\ t \geq 0\}$ form a strongly continuous contraction semi-group on C_0.

To conclude this section, we mention that it is worthwhile to study the regularity properties of the function $g(x) = E_x[f(X(t))]$. [33] contains some results for a certain class of functions f. However, for more general functions, e.g., Lip-1 functions, the questions of whether g is H-differentiable, whether $D^2 g(x)$ is of Hilbert-Schmidt type and whether $D^2 g(x)$ is of trace class type are still open.

§ 6. Divergence theorem.

The formulation of divergence theorem in an oriented n-dimensional Riemannian manifold depends on the volume element which in turn depends on the Lebesgue measure in $\mathbb{R}^n$. Note that the Lebesgue measure has no analogue in infinite dimensional space as we showed in the very beginning of these notes. However, we can use the abstract Wiener measure $p_t(x, dy)$ in an abstract Wiener space (i, H, B) to formulate a version of divergence theorem, which reduces to the usual one when B is finite dimensional. This formulation of divergence theorem is due to Goodman [15].

Definition 6.1. A real-valued function f defined on an open subset U of B is said to be H-C^1 if (i) f is continuous and H-differentiable and (ii) Df is continuous from U into H and $Df(x) \in B^*$ for all x in U.

Theorem 6.1. Let $\{U_\alpha;\ \alpha \in A\}$ be an open covering of B. Then there exists an H-C^1 partition of unity $\{f_n;\ n=1,2,\cdots\}$ such that the support of each f_n is contained in some U_α.

Definition 6.2. A subset S of B is called an H-C^1 surface if for each x in S there is an open neighborhood U of x and an H-C^1 function f defined in U such that $Df(x) \neq 0$ and $S \cap U = \{y \in U;\ f(y) = 0\}$.

Definition 6.3. A subset V of B is said to have an H-C^1 boundary if for each point x in the boundary ∂V of V there is a neighborhood U of x and an H-C^1 function f defined in U such that $Df(x) \neq 0$ and $V \cap U = \{y \in U; f(y) < 0\}$.

Theorem 6.2. For any $0 < \alpha < \beta$ there exists a set V with an H-C^1 boundary such that $\{\|x\| < \alpha\} \subset V \subset \{\|x\| < \beta\}$.

Remark. The theorem is trivial if $\|x\|$ is an H-C^1 function off the origin.

Proof. Take a number $c > 0$ such that $c < \alpha < \beta < c^{-1}$. It is easy to see that the function $f(x) = \min\{\|x\|, 2c^{-1}\}$ is a bounded Lip-1 function. Hence by the proof of Theorem 4.6 (a) and Chapter II Theorem 6.2 $g_t = p_t f$ is an H-C^1 function and by Chapter II Theorem 6.1 g_t converges to f uniformly as $t \to 0$. Let $r = (\alpha + \beta)/2$. Then for sufficiently small t, the set $V_t = \{g_t < r\}$ satisfies $\{\|x\| < \alpha\} \subset V_t \subset \{\|x\| < \beta\}$. We show that V_t has an H-C^1 boundary for small t. To do this, it is sufficient to show that Dg_t is nonzero on the set $\{c < \|x\| < c^{-1}\}$ for small t.

Let x be a fixed vector in $\{c < \|x\| < c^{-1}\}$. Pick $h \in H$ such that $\|h - x\| < \|x\|/4$. Then $\|h\| < 5\|x\|/4$. Hence whenever $0 < \varepsilon < \frac{1}{2}$ and $\|y\| < \|x\|/4$, we have $\|x+\varepsilon h+y\| < 2c^{-1}$. Therefore,

$$f(x + \varepsilon h + y) = \|x + \varepsilon h + y\| .$$

Moreover, when $\|y\| < \|x\|/_4$, we also have $\|x + y\| < 2c^{-1}$, so

$$f(x + y) = \|x + y\|.$$

Now, when $\|y\| < \|x\|_{/4}$, we have

$$\|x+\varepsilon h+y\| - \|x+y\| = \|(1+\varepsilon)(x+y)+\varepsilon(h-x-y)\| - \|x+y\|$$

$$\geq (1+\varepsilon)\|x+y\| - \varepsilon\|h-x-y\| - \|x+y\|$$

$$= \varepsilon[\|x+y\| - \|h-x-y\|].$$

But $\|x+y\| > \|x\| - \|y\| > \frac{3}{4}\|x\|$,

and $\|h-x-y\| \leq \|h-x\| + \|y\| < \frac{\|x\|}{2}$.

Therefore,

$$\|x+\varepsilon h+y\| - \|x+y\| \geq \frac{\varepsilon}{4}\|x\| \geq \frac{\varepsilon}{4}\frac{4}{5}\|h\| = \frac{\varepsilon}{5}\|h\|.$$

Thus we have shown that when $0 < \varepsilon < \frac{1}{2}$,

$$\int_{\|y\| < \|x\|/4} (f(x+\varepsilon h+y) - f(x+y))\,p_t(dy) > \frac{\varepsilon}{5}\|h\|\,p_t\{\|y\| < \frac{\|x\|}{4}\}.$$

On the other hand, since $|f(u)-f(v)| \leq \|u-v\|$ for all u and v in B, we have

$$\int_{\|y\| \geq \frac{\|x\|}{4}} |f(x+\varepsilon h+y) - f(x+y)|\,p_t(dy) \leq \varepsilon\|h\|\,p_t\{\|y\| \geq \frac{\|x\|}{4}\}.$$

Hence

$$\frac{1}{\varepsilon}[g_t(x+\varepsilon h)-g_t(x)]=\frac{1}{\varepsilon}\int_B (f(x+\varepsilon h+y)-f(x+y))p_t(dy)$$

$$\geq \varepsilon\|h\|\,[\tfrac{1}{5}\,p_t\{\|y\|<\|x\|/4\}-p_t\{\|y\|\geq\|x\|/4\}].$$

Letting $\varepsilon \to 0$, we obtain

$$(Dg_t(x),h)\geq \|h\|\,[\tfrac{1}{5}p_t\{\|y\|<\|x\|/4\}-p_t\{\|y\|\geq\|x\|/4\}].$$

Therefore

$$\|Dg_t(x)\|_* \geq \tfrac{1}{5}\,p_t\{\|y\|<\|x\|/4\}-p_t\{\|y\|\geq\|x\|/4\}$$

$$\geq \tfrac{1}{5}\,p_t\{\|y\|<c/4\}-p_t\{\|y\|\geq c/4\}.$$

Observe that $p_t\{\|y\|<\frac{c}{4}\}\to 1$ as $t\to 0$. Choose $\delta>0$ such that $p_\delta\{\|y\|<\frac{c}{4}\}>\frac{6}{7}$. Then

$$\|Dg_\delta(x)\|_* \geq \frac{1}{35}$$

for all x in $\{c<\|x\|<c^{-1}\}$. #

<u>Definition 6.4</u>. Let M be a differentiable surface in H. A 1-dimensional orthogonal projection P of H is called a <u>normal projection</u> for M at $x \in M$ if

$$|P(y-x)|=o(|y-x|),\quad y\in M \text{ and } |y-x|\to 0.$$

<u>Theorem 6.3.</u> Let S ba an H-C^1 surface in (i, H, B). Then there exists a unique map $N : S \to L(B, B^*)$ such that

(a) for each $y \in S$, $N_y|_H$ is a normal projection for $(S-y) \cap H$ at 0,

(b) for each $y \in S$, $J_y = I - N_y$ is a homeomorphism from an open neighborhood of y in S onto an open subset of ker N_y,

(c) $N : S \to L(H)$ (with operator norm topology) is continuous.

Let S be an H-C^1 surface. It follows from the above theorem that for each $y \in S$ there is $h \in B^*$, $|h| = 1$ and an open neighborhood U of y in S such that (i) $N_y h = h$, (ii) $N_w h \neq 0$ for all $w \in U$ and (iii) $J_y = I - N_y$ is a homeomorphism of U onto $J_y(U) \subset$ ker N_y. h is called a <u>unit normal</u> at y and U is called a <u>coordinate neighborhood</u> of y. It is easy to see that there is a unique unit normal n(y) such that $y - \varepsilon n(y) \in U$ for small $\varepsilon > 0$. n(y) is called the <u>outward normal</u> at y.

<u>Theorem 6.4.</u> (Existence of surface measure). Let S be an H-C^1 surface in (i, H, B). For $x \in B$ and $t > 0$ there exists a unique Borel measure $\sigma_t(x, \cdot)$ on S such that for any coordinate neighborhood U of $y \in S$ and any Borel subset E of U, there holds

$$\sigma_t(x, E) = \frac{1}{\sqrt{2\pi t}} \int_{J_y(E)} |N_{J_y^{-1} z} h|^{-1} \exp[-|N_y(J_y^{-1} z - x)|^2/2t] p_t'(J_y x, dz),$$

where h is a unit normal at y and $p_t'(\cdot)$ is the Wiener measure in ker N_y with parameter t. $\sigma_t(x, \cdot)$ is called the normal surface measure on S with dilation parameter t and translation variable x.

Definition 6.5. Let f be a measurable function from a subset U of B into B*. Suppose f is H-differentiable and $Df(y) \in \mathcal{L}_{(1)}(H)$ for all $y \in U$. For each $x \in B$ and $t > 0$ we define the divergence of f at x by

$$\mathrm{div}_{t,x} f(y) = \mathrm{trace}\, Df(y) - \frac{1}{t}(f(y), y-x), \quad y \in U.$$

Theorem 6.5. (Divergence theorem). Let V be a subset of B with an H-C^1 boundary. Let $g : V \cup \partial V \to H$ be a measurable function satisfying the following conditions:

(a) g is H-differentiable in V and H-continuous in ∂V,

(b) $Dg : V \to \mathcal{L}(H)$ (with the weak operator topology) is continuous,

(c) for each $x \in B$, $t > 0$ and $h \in H$, the functions $\langle g(\cdot), h\rangle^2$ and $|Dg(\cdot)h|$ are $p_t(x, dy)$-integrable on V and the function $|g(\cdot)|$ is $\sigma_t(x, dy)$-integrable on ∂V.

If T is a test operator then

$$\int_V \mathrm{div}_{t,x} Tg(y) p_t(x,dy) = \int_{\partial V} \langle Tg(y), n(y)\rangle\, \sigma_t(x, dy),$$

where $\sigma_t(x,\cdot)$ is the normal surface measure on ∂V with dilation parameter t and translation variable x.

Corollary 6.1. Let V be a subset of B with an H-C^1 boundary. Let $f : V \cup \partial V \to H$ be a measurable function satisfying the following conditions:

(a) f is H-differentiable in V and H-continuous in ∂V,

(b) $f(x) \in B^*$ for all x in V and $\sup_{x \in V} \| f(x) \|_* < \infty$,

(c) $Df(x) \in \mathcal{L}_{(1)}(H)$ for all x in V and Df is continuous from V into $\mathcal{L}(H)$ (with the weak operator topology),

(d) for each $x \in B$ and $t > 0$, $|f(\cdot)|$ is $\sigma_t(x, dy)$-integrable on ∂V and $\| Df(\cdot) \|_1$ is $p_t(x, dy)$-integrable on V.

Then

$$\int_V \mathrm{div}_{t,x} f(y) p_t(x, dy) = \int_{\partial V} \langle f(y), n(y) \rangle \sigma_t(x, dy).$$

One may attempt to obtain a divergence theorem for infinite dimensional manifold. Just as Lebesgue measure has no analogue in ∞-dim space, volume element has no analogue in ∞-dim manifold. Nevertheless, one can formulate a version of divergence theorem for ∞-dim manifold without using the non-existent volume element. We describe briefly a possible approach as follows. Let W be a Riemann-Wiener manifold [30]. W is a Banach manifold modelled on (i, H, B). Each tangent space of W is equipped with a norm and a densely defined inner product. By using stochastic integral (§5), a Brownian motion $B(t)$ is constructed on W in [32]. Let $r_t(x, E) = \mathrm{Prob}\{B(t) \in E \mid B(0) = x\}$, $t > 0$, $x \in W$ and $E \in \mathcal{B}(W)$. $r_t(x, dy)$ will take the place of $p_t(x, dy)$ in the formulation of divergence theorem. The notion of H-C^1 surface and H-C^1 boundary can be generalized to W in an obvious way. However, the construction of surface measure corresponding to $r_t(x, dy)$ is by no means easy and remains unsolved.

A generalization of Theorem 6.5 and Corollary 6.1 to non-Gaussian measure is of interest. Let $X(t)$ be the solution of the stochastic integral equation in Theorem 5.6 and let $q_t(x, dy) = \text{Prob}\{X(t) \in dy | X(0) = x\}$. The construction of surface measure corresponding to $q_t(x, dy)$ is still open and the formulation of divergence theorem is unknown so far.

§7. Comments on Chapter III.

§1. In the proof of Theorem 1.1 if we choose another summable sequence $\{\lambda_n\}$ of positive numbers and another countable dense subset $\{a_n\}$ of B, we will have a different Hilbert space $\tilde{H}$. However, by Chapter II Theorem 5.3, we will have the same Hilbert space H. Note that Kuelbs' method does not give a constructive way to describe H. The other proo sketched in this section yields a characterization of thos elements in B belonging to H, namely, $H=\{y \in B;\ S_y 1 \in \mathcal{L}_0\}$. For instance, when $B=C[0, 1]$ and μ is the Wiener measure w, then $S_y 1(x)= \int_0^1 y'(t)dx(t)$, $x \in C[0, 1]$, i.e., y is absolutely continuous and $y' \in L^2[0, 1]$.

§2. See [27] for the relation between the abstract Wiener spaces and the reproducing kernel Hilbert spaces.

§3. In some of the theorems in [19; 30; 37] the p_1-integrabil of $\|\cdot\|^r$ for various $r \geq 1$ is assumed. It follows from Fernique's theorem or Skorokhod's theorem that this assum tion is superfluous.

§4. Our method in proving Theorem 4.6 and Theorem 4.7 is different from Gross' [19]. Our estimates in the proofs depend crucially on Lemma 4.3 and Lemma 4.4 and the computations are simpler than Gross'.

§5. In developing stochastic integrals on an abstract Wiener space [31; 32; 33; 35] we have made the following assumption on (i, H, B): there exists an increasing sequence $\{P_n\}$ of finite dimensional projections on B such that $P_n(B) \subset B^*$, P_n converges strongly to the identity on B and $P_n|_H$ converges strongly to the identity on H. This assumption is unnecessary for our previous papers concerning stochastic integrals. The existence of (i, H, B_0) and $\{Q_n\}$ stated in the beginning of this section is good enough.

§6. Gross has pointed out in [19] that analysis over infinite dimensional manifolds is a potentially rich field for investigation. The divergence theorem is one step toward the study of cohomology theory over Riemann-Wiener manifolds.

References

1. Bonic R. and Frampton J., Differentiable functions on certain Banach spaces, Bull. Amer. Math. Soc. 71(1965), 393-395.
2. Cameron R. H. and Graves R., Additive functionals on a space of continuous funcitons, Trans. Amer. Math. Soc. 70 (1951), 160-176.
3. Cameron R. H. and Martin W. T., Transformation of Wiener integrals under translations, Ann. Math. 45(1944), 386-396.
4. __________, Transformation of Wiener integrals under a general class of linear transformations, Trans. Amer. Math. Soc. 58(1945), 184-219.
5. __________, The transformation of Wiener integrals by non-linear transformations, Trans. Amer. Math. Soc. 66(1949), 253-283.
6. Donsker M. D. and Lions J. L., Volterra variational equation boundary value problems and function space integrals, Acta Math. 108(1962), 147-228.
7. Donsker M. D., On function space integrals, in "Analysis in function space edited by W.T. Martin and I.E. Segal (1963), 17-30.
8. __________, Lecture notes in "Integration in function space' Courant Institute, New York University (1971).
9. Dudley R.M., Feldman J. and LeCam L., On semi-norms and probabilities, and abstract Wiener spaces., Ann. Math 93

(1971), 390-408.

10. Feldman J., Equivalence and perpendicularity of Gaussian processes, Pacific J. Math. 8(1958), 699-708.

11. __________, A short proof of the Levy continuity theorem in Hilbert space, Israel J. Math. 3 (1965), 99-103.

12. Fernique M.X., Intégrabilité des Vecteurs Gaussiens, Academie des Sciences, Paris, Comptes Rendus, 270 Séries A (1970), 1698-1699.

13. Gel'fand I.M. and Vilenkin N.Ya., Generalized functions, vol. 4, English translation, Academic Press, New York(1964).

14. Gihman I.I. and Skorokhod A.V., Densities of probability measures in function space, Russian Math. Surveys (Engl. transl.) 21 (1966), 83-156.

15. Goodman V., A divergence theorem for Hilbert space, Trans. Amer. Math. Soc. 164 (1972), 411-426.

16. Gross L., Measurable functions on Hilbert space, Trans. Amer. Math. Soc. 105(1962), 372-390.

17. __________, Harmonic analysis on Hilbert space, Memoirs Amer. Math.Soc. no. 46(1963).

18. __________, Abstract Wiener spaces, Proc. 5th. Berkeley Sym. Math. Stat. Prob. 2(1965), 31-42.

19. __________, Potential theory on Hilbert space, J. Func. Anal. 1(1967), 123-181.

20. Hajek J., A property of J-divergences of marginal probability distributions, Czech. Math. J. 8(1958), 460-463.

21. __________, On a property of normal distributions of an

arbitrary stochastic process, Czech. Math. J. 8(1958), 610-618.

22. Helms L.L., Mean convergence of martingales, Trans. Amer. Math. Soc. 87(1958), 439-446.

23. Ito K., The topological support of Gauss measures on Hilbert space, Nagoya Math. J. 38(1970), 181-183.

24. ________, Lecture notes in "Stochastic integrals", Cornell University (1972).

25. Kac M., On distributions of certain Wiener integrals, Trans. Amer. Math. Soc. 65(1949), 1-13.

26. Kakutani S., On equivalence of infinite product measures, Ann. Math. 49(1948), 214-224.

27. Kallianpur G., Abstract Wiener processes and their reproducing kernel Hilbert spaces, Z. Wahrscheinlichkeitstheorie 17(1971), 113-123.

28. Kato T., Perturbation theory for linear oeprators, Springer-Verlag, Berlin and New York (1966).

29. Kuelbs J., Gaussian measures on a Banach space J. Func. Anal. 5(1970), 354-367.

30. Kuo H.-H., Integration theory on infinite-dimensional manifolds, Trans. Amer. Math. Soc. 159(1971), 57-78.

31. ________, Stochastic integrals in abstract Wiener space, Pacific J. Math. 41(1972), 469-483.

32. ________, Diffusion and Brownian motion on infinite-dimensional manifolds, Trans. Amer. Math. Soc. 169(1972), 439-457.

33. __________, Stochastic integrals in abstract Wiener space II :Regularity properties, Nagoya Math. J. 50(1973), 89-116.

34. __________, Integration by parts for abstract Wiener measures, Duke Math. J. 41(1974), 373-379.

35. Kuo H.-H. and Piech M.A., Stochastic integral and parabolic equation in abstract Wiener space, Bull. Amer. Math. Soc. 79(1973), 478-482.

36. Maruyama G., Notes on Wiener integrals, Kodai Math. Seminar Rep. 3(1950), 41-44.

37. Piech M.A., A fundamental solution of the parabolic equation on Hilbert space, J. Func. Anal. 3(1969), 85-114.

38. Prohorov Yu. V., Convergence for random processes and limit theorems in probability theory, Teor. Veroj. i Prim. 1(1956), 177-238.

39. __________, The method of characteristic functionals, Proc. 4th Berkeley Sym. Math. Stat. Prob. (1961), 403-419.

40. Sazonov V. V., A remark on characteristic functionals, Teor. Veroj. i Prim. 3(1958), 201-205.

41. Segal I.E., Tensor algebras over Hilbert spaces, Trans. Amer. Math. Soc. 81(1956), 106-134.

42. __________, Distributions in Hilbert space and canonical systems of operators, Trans. Amer. Math. Soc. 88(1958),12-41.

43. Shepp L.A., Gaussian measures in function spaces, Pacific J. Math. 17(1966), 167-173.

44. Skorokhod A.V., Notes on Gaussian measures in a Banach space, Teor. Veroj. i Prim. 15(1970), 519-520.

45. Sunouchi G., Harmonic analysis and Wiener integrals, Tohoku Math. J. 3(1951), 187-196.

46. Varadhan S.R.S., Stochastic processes, Courant Institute, New York University (1968).

47. Whitfield J.H.M., Differentiable functions with bounded non-empty support on Banach spaces, Bull. Amer. Math. Soc. 72 (1965), 145-146.

48. Wiener N., The average value of a functional, Proc. London Math. Soc. 22(1922), 454-467.

49. ________, Differential space, J. Math. Phys. 58(1923), 131-174.

50. Selected Papers of Nobert Wiener, SIAM and MIT Press (1965).

51. Yeh J., Stochastic processes and the Wiener integral, Marcel Dekker, New York (1973).

INDEX

309: D. H. Sattinger, Topics in Stability and Bifurcation Theory. 90 pages. 1973. DM 20,-

310: B. Iversen, Generic Local Structure of the Morphisms in mutative Algebra. IV, 108 pages. 1973. DM 18,-

311: Conference on Commutative Algebra. Edited by J. W. Brewer E. A. Rutter. VII, 251 pages. 1973. DM 24,-

312: Symposium on Ordinary Differential Equations. Edited by Harris, Jr. and Y. Sibuya. VIII, 204 pages. 1973. DM 22,-

313: K. Jörgens and J. Weidmann, Spectral Properties of Hamiln Operators. III, 140 pages. 1973. DM 18,-

14: M. Deuring, Lectures on the Theory of Algebraic Functions e Variable. VI, 151 pages. 1973. DM 18,-

15: K. Bichteler, Integration Theory (with Special Attention to Measures). VI, 357 pages. 1973. DM 29,-

16: Symposium on Non-Well-Posed Problems and Logarithmic xity. Edited by R. J. Knops. V, 176 pages. 1973. DM 20,-

17: Séminaire Bourbaki - vol. 1971/72. Exposés 400-417. IV, ages. 1973. DM 29,-

18: Recent Advances in Topological Dynamics. Edited by A. VIII, 285 pages. 1973. DM 27,-

9: Conference on Group Theory. Edited by R. W. Gatterdam W. Weston. V, 188 pages. 1973. DM 20,-

20: Modular Functions of One Variable I. Edited by W. Kuyk. V, ges. 1973. DM 20,-

1: Séminaire de Probabilités VII. Edité par P. A. Meyer. VI, 322 1973. DM 29,-

2: Nonlinear Problems in the Physical Sciences and Biology. by I. Stakgold, D. D. Joseph and D. H. Sattinger. VIII, 357 1973. DM 29,-

3: J. L. Lions, Perturbations Singulières dans les Problèmes ites et en Contrôle Optimal. XII, 645 pages. 1973. DM 46,-

: K. Kreith, Oscillation Theory. VI, 109 pages. 1973. DM 18,-

5: C.-C. Chou, La Transformation de Fourier Complexe et ion de Convolution. IX, 137 pages. 1973. DM 18,-

: A. Robert, Elliptic Curves. VIII, 264 pages. 1973. DM 24,-

7: E. Matlis, One-Dimensional Cohen-Macaulay Rings. XII, es. 1973. DM 20,-

: J. R. Büchi and D. Siefkes, The Monadic Second Order f All Countable Ordinals. VI, 217 pages. 1973. DM 22,-

W. Trebels, Multipliers for (C, α)-Bounded Fourier Expan-Banach Spaces and Approximation Theory. VII, 103 pages. M 18,-

Proceedings of the Second Japan-USSR Symposium on ty Theory. Edited by G. Maruyama and Yu. V. Prokhorov. ages. 1973. DM 40,-

Summer School on Topological Vector Spaces. Edited by roeck. VI, 226 pages. 1973. DM 22,-

Séminaire Pierre Lelong (Analyse) Année 1971-1972. V, 131 73. DM 18,-

Numerische, insbesondere approximationstheoretische Be-von Funktionalgleichungen. Herausgegeben von R. An-W. Törnig. VI, 296 Seiten. 1973. DM 27,-

F. Schweiger, The Metrical Theory of Jacobi-Perron Algo- pages. 1973. DM 18,-

H. Huck, R. Roitzsch, U. Simon, W. Vortisch, R. Walden, B. d W. Wendland, Beweismethoden der Differentialgeome-Ben. IX, 159 Seiten. 1973. DM 20,-

L'Analyse Harmonique dans le Domaine Complexe. Edité utowicz. VIII, 169 pages. 1973. DM 20,-

ambridge Summer School in Mathematical Logic. Edited Mathias and H. Rogers. IX, 660 pages. 1973. DM 46,-

Lindenstrauss and L. Tzafriri, Classical Banach Spaces. es. 1973. DM 24,-

G. Kempf, F. Knudsen, D. Mumford and B. Saint-Donat, mbeddings I. VIII, 209 pages. 1973. DM 22,-

Groupes de Monodromie en Géométrie Algébrique. Par P. Deligne et N. Katz. X, 438 pages. 1973. DM 44,-

Algebraic K-Theory I, Higher K-Theories. Edited by , 335 pages. 1973. DM 29,-

Vol. 342: Algebraic K-Theory II, "Classical" Algebraic K-Theory, and Connections with Arithmetic. Edited by H. Bass. XV, 527 pages. 1973. DM 40,-

Vol. 343: Algebraic K-Theory III, Hermitian K-Theory and Geometric Applications. Edited by H. Bass. XV, 572 pages. 1973. DM 40,-

Vol. 344: A. S. Troelstra (Editor), Metamathematical Investigation of Intuitionistic Arithmetic and Analysis. XVII, 485 pages. 1973. DM 38,-

Vol. 345: Proceedings of a Conference on Operator Theory. Edited by P. A. Fillmore. VI, 228 pages. 1973. DM 22,-

Vol. 346: Fučík et al., Spectral Analysis of Nonlinear Operators. II, 287 pages. 1973. DM 26,-

Vol. 347: J. M. Boardman and R. M. Vogt, Homotopy Invariant Algebraic Structures on Topological Spaces. X, 257 pages. 1973. DM 24,-

Vol. 348: A. M. Mathai and R. K. Saxena, Generalized Hypergeometric Functions with Applications in Statistics and Physical Sciences. VII, 314 pages. 1973. DM 26,-

Vol. 349: Modular Functions of One Variable II. Edited by W. Kuyk and P. Deligne. V, 598 pages. 1973. DM 38,-

Vol. 350: Modular Functions of One Variable III. Edited by W. Kuyk and J.-P. Serre. V, 350 pages. 1973. DM 26,-

Vol. 351: H. Tachikawa, Quasi-Frobenius Rings and Generalizations. XI, 172 pages. 1973. DM 20,-

Vol. 352: J. D. Fay, Theta Functions on Riemann Surfaces. V, 137 pages. 1973. DM 18,-

Vol. 353: Proceedings of the Conference on Orders, Group Rings and Related Topics. Organized by J. S. Hsia, M. L. Madan and T. G. Ralley. X, 224 pages. 1973. DM 22,-

Vol. 354: K. J. Devlin, Aspects of Constructibility. XII, 240 pages. 1973. DM 24,-

Vol. 355: M. Sion, A Theory of Semigroup Valued Measures. V, 140 pages. 1973. DM 18,-

Vol. 356: W. L. J. van der Kallen, Infinitesimally Central-Extensions of Chevalley Groups. VII, 147 pages. 1973. DM 18,-

Vol. 357: W. Borho, P. Gabriel und R. Rentschler, Primideale in Einhüllenden auflösbarer Lie-Algebren. V, 182 Seiten. 1973. DM 20,-

Vol. 358: F. L. Williams, Tensor Products of Principal Series Representations. VI, 132 pages. 1973. DM 18,-

Vol. 359: U. Stammbach, Homology in Group Theory. VIII, 183 pages. 1973. DM 20,-

Vol. 360: W. J. Padgett and R. L. Taylor, Laws of Large Numbers for Normed Linear Spaces and Certain Fréchet Spaces. VI, 111 pages. 1973. DM 18,-

Vol. 361: J. W. Schutz, Foundations of Special Relativity: Kinematic Axioms for Minkowski Space Time. XX, 314 pages. 1973. DM 26,-

Vol. 362: Proceedings of the Conference on Numerical Solution of Ordinary Differential Equations. Edited by D. Bettis. VIII, 490 pages. 1974. DM 34,-

Vol. 363: Conference on the Numerical Solution of Differential Equations. Edited by G. A. Watson. IX, 221 pages. 1974. DM 20,-

Vol. 364: Proceedings on Infinite Dimensional Holomorphy. Edited by T. L. Hayden and T. J. Suffridge. VII, 212 pages. 1974. DM 20,-

Vol. 365: R. P. Gilbert, Constructive Methods for Elliptic Equations. VII, 397 pages. 1974. DM 26,-

Vol. 366: R. Steinberg, Conjugacy Classes in Algebraic Groups (Notes by V. V. Deodhar). VI, 159 pages. 1974. DM 18,-

Vol. 367: K. Langmann und W. Lütkebohmert, Cousinverteilungen und Fortsetzungssätze. VI, 151 Seiten. 1974. DM 16,-

Vol. 368: R. J. Milgram, Unstable Homotopy from the Stable Point of View. V, 109 pages. 1974. DM 16,-

Vol. 369: Victoria Symposium on Nonstandard Analysis. Edited by A. Hurd and P. Loeb. XVIII, 339 pages. 1974. DM 26,-

Vol. 370: B. Mazur and W. Messing, Universal Extensions and One Dimensional Crystalline Cohomology. VII, 134 pages. 1974. DM 16,–

Vol. 371: V. Poenaru, Analyse Différentielle. V, 228 pages. 1974. DM 20,–

Vol. 372: Proceedings of the Second International Conference on the Theory of Groups 1973. Edited by M. F. Newman. VII, 740 pages. 1974. DM 48,–

Vol. 373: A. E. R. Woodcock and T. Poston, A Geometrical Study of the Elementary Catastrophes. V, 257 pages. 1974. DM 22,–

Vol. 374: S. Yamamuro, Differential Calculus in Topological Linear Spaces. IV, 179 pages. 1974. DM 18,–

Vol. 375: Topology Conference 1973. Edited by R. F. Dickman Jr. and P. Fletcher. X, 283 pages. 1974. DM 24,–

Vol. 376: D. B. Osteyee and I. J. Good, Information, Weight of Evidence, the Singularity between Probability Measures and Signal Detection. XI, 156 pages. 1974. DM 16,–

Vol. 377: A. M. Fink, Almost Periodic Differential Equations. VIII, 336 pages. 1974. DM 26,–

Vol. 378: TOPO 72 – General Topology and its Applications. Proceedings 1972. Edited by R. Alò, R. W. Heath and J. Nagata. XIV, 651 pages. 1974. DM 50,–

Vol. 379: A. Badrikian et S. Chevet, Mesures Cylindriques, Espaces de Wiener et Fonctions Aléatoires Gaussiennes. X, 383 pages. 1974. DM 32,–

Vol. 380: M. Petrich, Rings and Semigroups. VIII, 182 pages. 1974. DM 18,–

Vol. 381: Séminaire de Probabilités VIII. Edité par P. A. Meyer. IX, 354 pages. 1974. DM 32,–

Vol. 382: J. H. van Lint, Combinatorial Theory Seminar Eindhoven University of Technology. VI, 131 pages. 1974. DM 18,–

Vol. 383: Séminaire Bourbaki – vol. 1972/73. Exposés 418-435 IV, 334 pages. 1974. DM 30,–

Vol. 384: Functional Analysis and Applications, Proceedings 1972. Edited by L. Nachbin. V, 270 pages. 1974. DM 22,–

Vol. 385: J. Douglas Jr. and T. Dupont, Collocation Methods for Parabolic Equations in a Single Space Variable (Based on C¹-Piecewise-Polynomial Spaces). V, 147 pages. 1974. DM 16,–

Vol. 386: J. Tits, Buildings of Spherical Type and Finite BN-Pairs. IX, 299 pages. 1974. DM 24,–

Vol. 387: C. P. Bruter, Eléments de la Théorie des Matroïdes. V, 138 pages. 1974. DM 18,–

Vol. 388: R. L. Lipsman, Group Representations. X, 166 pages. 1974. DM 20,–

Vol. 389: M.-A. Knus et M. Ojanguren, Théorie de la Descente et Algèbres d' Azumaya. IV, 163 pages. 1974. DM 20,–

Vol. 390: P. A. Meyer, P. Priouret et F. Spitzer, Ecole d'Eté de Probabilités de Saint-Flour III – 1973. Edité par A. Badrikian et P.-L. Hennequin. VIII, 189 pages. 1974. DM 20,–

Vol. 391: J. Gray, Formal Category Theory: Adjointness for 2-Categories. XII, 282 pages. 1974. DM 24,–

Vol. 392: Géométrie Différentielle, Colloque, Santiago de Compostela, Espagne 1972. Edité par E. Vidal. VI, 225 pages. 1974. DM 20,–

Vol. 393: G. Wassermann, Stability of Unfoldings. IX, 164 pages. 1974. DM 20,–

Vol. 394: W. M. Patterson 3rd, Iterative Methods for the Solution of a Linear Operator Equation in Hilbert Space – A Survey. III, 183 pages. 1974. DM 20,–

Vol. 395: Numerische Behandlung nichtlinearer Integrodifferential- und Differentialgleichungen. Tagung 1973. Herausgegeben von R. Ansorge und W. Törnig. VII, 313 Seiten. 1974. DM 28,–

Vol. 396: K. H. Hofmann, M. Mislove and A. Stralka, The Pontryagin Duality of Compact O-Dimensional Semilattices and its Applications. XVI, 122 pages. 1974. DM 18,–

Vol. 397: T. Yamada, The Schur Subgroup of the Brauer Group. V, 159 pages. 1974. DM 18,–

Vol. 398: Théories de l'Information, Actes des Rencontres de Marseille-Luminy, 1973. Edité par J. Kampé de Fériet et C. Picard. XII, 201 pages. 1974. DM 23,–

Vol. 399: Functional Analysis and its Applications, Proceedin 1973. Edited by H. G. Garnir, K. R. Unni and J. H. Williamsc XVII, 569 pages. 1974. DM 44,–

Vol. 400: A Crash Course on Kleinian Groups – San Francis 1974. Edited by L. Bers and I. Kra. VII, 130 pages. 1974. DM 18

Vol. 401: F. Atiyah, Elliptic Operators and Compact Grou V, 93 pages. 1974. DM 18,–

Vol. 402: M. Waldschmidt, Nombres Transcendants. VIII, 2 pages. 1974. DM 25,–

Vol. 403: Combinatorial Mathematics – Proceedings 1972. Edi by D. A. Holton. VIII, 148 pages. 1974. DM 18,–

Vol. 404: Théorie du Potentiel et Analyse Harmonique. Edité J. Faraut. V, 245 pages. 1974. DM 25,–

Vol. 405: K. Devlin and H. Johnsbråten, The Souslin Probl VIII, 132 pages. 1974. DM 18,–

Vol. 406: Graphs and Combinatorics – Proceedings 1973. Ed by R. A. Bari and F. Harary. VIII, 355 pages. 1974. DM

Vol. 407: P. Berthelot, Cohomologie Cristalline des Schéma Caracteristique p > o. VIII, 598 pages. 1974. DM 44,–

Vol. 408: J. Wermer, Potential Theory. VIII, 146 pages. 1 DM 18,–

Vol. 409: Fonctions de Plusieurs Variables Complexes, Sémir François Norguet 1970–1973. XIII, 612 pages. 1974. DM

Vol. 410: Séminaire Pierre Lelong (Analyse) Année 1972–1 VI, 181 pages. 1974. DM 18,–

Vol. 411: Hypergraph Seminar. Ohio State University, 1 Edited by C. Berge and D. Ray-Chaudhuri. IX, 287 pages. 1 DM 28,–

Vol. 412: Classification of Algebraic Varieties and Co Complex Manifolds. Proceedings 1974. Edited by H. Pop 333 pages. 1974. DM 30,–

Vol. 413: M. Bruneau, Variation Totale d'une Fonction. XIV pages. 1974. DM 30,–

Vol. 414: T. Kambayashi, M. Miyanishi and M. Takeuchi potent Algebraic Groups. VI, 165 pages. 1974. DM 20,–

Vol. 415: Ordinary and Partial Differential Equations, Procee of the Conference held at Dundee, 1974. XVII, 447 pages. DM 37,–

Vol. 416: M. E. Taylor, Pseudo Differential Operators. I pages. 1974. DM 18,–

Vol. 417: H. H. Keller, Differential Calculus in Locally Spaces. XVI, 131 pages. 1974. DM 18,–

Vol. 418: Localization in Group Theory and Homotopy and Related Topics Battelle Seattle 1974 Seminar. Ed P. J. Hilton. VI, 171 pages. 1974. DM 20,–

Vol. 419: Topics in Analysis – Proceedings 1970. Ed O. E. Lehto, I. S. Louhivaara, and R. H. Nevanlinna. XIII, 391 1974. DM 35,–

Vol. 420: Category Seminar. Proceedings, Sydney C Theory Seminar 1972/73. Edited by G. M. Kelly. VI, 375 1974. DM 32,–

Vol. 421: V. Poénaru, Groupes Discrets. VI, 216 pages DM 23,–

Vol. 422: J.-M. Lemaire, Algèbres Connexes et Homolo Espaces de Lacets. XIV, 133 pages. 1974. DM 23,–

Vol. 423: S. S. Abhyankar and A. M. Sathaye, Geometric of Algebraic Space Curves. XIV, 302 pages. 1974. D

Vol. 424: L. Weiss and J. Wolfowitz, Maximum Pr Estimators and Related Topics. V, 106 pages. 1974. D

Vol. 425: P. R. Chernoff and J. E. Marsden, Properties o Dimensional Hamiltonian Systems. IV, 160 pages. 1974.

Vol. 426: M. L. Silverstein, Symmetric Markov Proce 287 pages. 1974. DM 28,–

Vol. 427: H. Omori, Infinite Dimensional Lie Transf Groups. XII, 149 pages. 1974. DM 18,–

Vol. 428: Algebraic and Geometrical Methods in Topol ceedings 1973. Edited by L. F. McAuley. XI, 280 pag DM 28,–